MINI Building

14 NEIGHBORHOOD FACILITY

CONTENTS

HP OFFICE BUILDING ｜ HP 오피스 빌딩	**004**
UNITEX HEADQUARTER BUILDING ｜ 유니텍스 사옥	**014**
BUILDING OF MUSIC ｜ 음악의 빌딩	**026**
NATIONAL CENTRE FOR ART, CRAFTS & DESIGN ｜ 국립미술공예디자인센터	**034**
VERTICAL STONE ｜ 세로돌	**046**
ESCOLA GRÉMIO ｜ 그레미오 학교	**058**
LA RINASCENTE – PIAZZA FIUME ｜ 라 리나센테 – 피우메 광장	**068**
ENHAKO BUILDING ｜ 엔하코 빌딩	**078**
URBAN GARDEN ｜ 도시 정원	**088**
MASJID CAHAYA JAMI AL- HURRIYAH ｜ 자미알 후리야 빛의 사원	**098**
BALTASAR BUILDING ｜ 발타사르 빌딩	**110**
GREYHOUND HOUSE ｜ 그레이하운드 하우스	**122**
KUBOMI APARTMENT ｜ 쿠보미 아파트먼트	**136**
HS77 ｜ 에이치에스77	**144**
BSP20 HOUSE ｜ 비에스피20 하우스	**152**

TBD OFFICE ǀ 티비디 사옥	**164**
DOZEN DOORS ǀ 더즌 도어스	**172**
TAIPEI PERFORMING ARTS CENTER ǀ 타이베이 공연예술센터	**180**
STEPPING HOUSE ǀ 높은계단집	**194**
EARLY BKK CAFE ǀ 얼리 비케이케이	**204**
STREET AND GARDEN APARTMENTS ǀ 스트리트 앤 가든 아파트	**210**
A JAPANESE MANGA ARTIST'S HOUSE ǀ 만화가의 집	**218**
OFFICE OF SOIL WASTE MANAGEMENT ǀ 토양폐기물 관리오피스	**226**
PLAZA_CIRCLE ǀ 원형광장_서클	**234**
PU-HAI PROPERTIES BUILDING ǀ 푸하이 부동산 빌딩	**242**
DARWIN 1111 ǀ 다윈 1111	**252**
RH+ BUILDING ǀ 알에이치플러스 빌딩	**260**
INDUSTRIALIZED HOUSE rNrH ǀ 산업화 하우스 rNrH	**268**
PROFILE ǀ 프로필	**280**

HP 오피스 빌딩
HP OFFICE BUILDING

ARCHITECT : NDT ARCHITECTURE / NGUYEN DANG TUONG

HP BUILDING IS LOCATED IN A DENSELY POPULATED AREA in Bac Ninh province. The project has the main facade to the northwest. The house has the west-facing facade and is situated close to the main road, we need to find a solution that not only helps create a unique facade but also ensures that the interior space is not affected by direct natural sunlight, the visibility is expanded and the dust and noise from surrounding vehicles are minimized. The main functions of the house include vaccination center (1st floor), a spa (2nd floor), rooms for rent (3rd, 4th, 5th floor), a cafe (6th floor). The facade is made up of hundreds of raised garden beds arranged at different heights, creating a giant layer of greenery. This solution helps to create a uniqueness for the house with better distribution of natural light to the interior space, better visibility and more effective prevention of dust and noise from outside.

A large void in the main lobby is directly linked to the atrium in the middle of the house, which helps bring a comfortable feeling when people step inside. The void and the atrium are directly connected to the stairs and the elevator to help people easily observe and move. In addition, this is also a solution to help ventilate and get better natural light for the middle area of the house. People moving between floors can feel the greenery and natural light everywhere. The 6th floor is used as an outdoor garden to create buffer spaces, which helps improve the quality of the indoor space. Large trees are used here to cool the indoor space, limit noise and allow natural ventilation into the house.

HP Building is a combination of practical solutions to overcome the limitations of the west-facing house, in which we focus on creating more resting spaces, increasing ventilation and natural light, improving the quality of indoor living space and inspiring green living to people in the area.

Location Bac Ninh Province, Vietnam **Use** Office **Site area** 440m² **Built area** 440m² **Gross area** 3,000m² **Completion** Feburury 2022 **Project manager** Nguyen Dang Tuong **Design team** Hoang Tuan Anh, Nguyen Van Tan, Ngo Thi Nga **Contractor** Hien construction **Photographer** Hoang Le

HP 빌딩은 박닌시의 인구 밀집 지역에 있다. 이 프로젝트는 북서쪽에 주 파사드가 있다. 소유자는 자연과 밀접하게 연결되고 자연 채광을 최대한 활용하며 충분한 휴식 공간이 있는 생활 공간을 요청했다. 또한 독특하고 눈에 띄는 파사드는 소유자의 요구 사항 중 하나이다. 빌딩은 서향 파사드를 가지고 있고 주요 도로에 인접해 있기 때문에 독특한 파사드가 요구 될 뿐만 아니라 내부 공간이 직사광선, 가시성 등의 영향을 받지 않도록 하는 솔루션을 통해 주변 차량의 먼지와 소음을 최소화했다. 주택의 주요 기능은 예방접종 센터(1층), 스파(2층), 임대룸(3, 4, 5층), 카페(6층)이 있다. 파사드는 서로 다른 높이로 배열된 수백 개의 높은 정원 화단으로 구성되어 거대한 녹지 공간을 이룬다. 이 솔루션은 자연광을 내부 공간으로 더 잘 분배하고 더 나은 가시성을 제공하며 외부의 먼지와 소음을 더 효과적으로 방지하여 집의 고유성을 만드는 데 도움이 된다. 메인 로비의 큰 보이드는 집 중앙의 아트리움과 직접 연결되어 사람이 들어왔을 때 편안한 느낌을 준다. 보이드와 아트리움은 계단과 엘리베이터로 직접 연결되어 사람들이 쉽게 관찰하고 이동할 수 있다. 또한 환기를 돕고 집 중앙에 더 나은 자연 채광을 제공하는 솔루션이다. 층간 이동은 사방에서 녹지와 자연 채광을 느낄 수 있다. 6층은 야외정원으로 활용하며 완충공간을 조성하여 실내공간의 품격을 높였다. 큰 나무는 실내 공간을 식히고 소음을 제한하며 집안으로의 자연 환기를 위해 사용되고 있다.

HP빌딩은 서향 주택의 한계를 극복하기 위한 실용적인 솔루션의 조합으로, 더 많은 휴식 공간 조성, 환기 및 자연 채광 증가, 실내 생활 공간의 질 향상 및 녹색 생활 영감에 중점을 두고 있다.

← Exterior view   → Exterior night view

↑ South-facing facade   ← Facade view

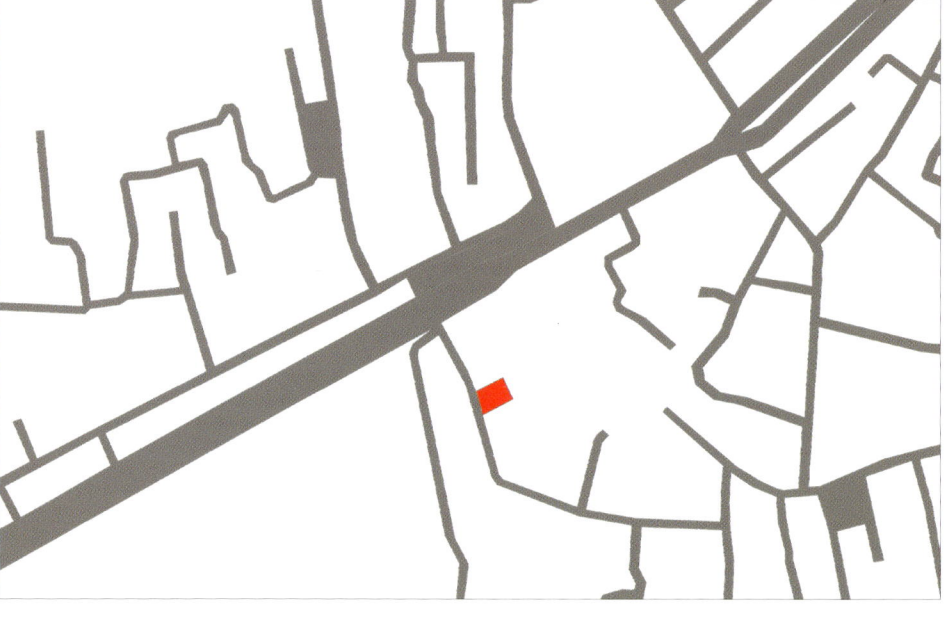

SITE PLAN

↑ Bird's eye view   ↳ View from south

NORTH ELEVATION

← The facade is creating a giant layer of greenery → The facade is made up of hundreds of raised garden

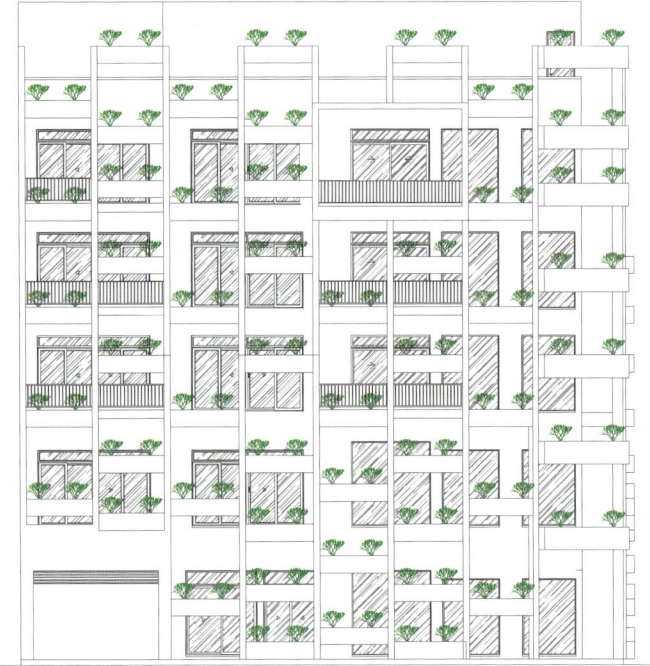 

WEST ELEVATION **SOUTH ELEVATION**

↖ Giant layer of greenery   ↑ Giant layer of greenery   ↗ Giant layer of greenery

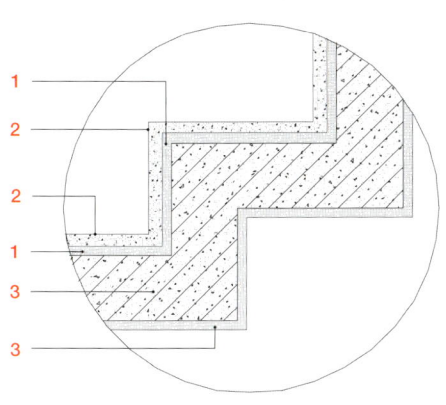

DETAIL ZIGSAG STAIR

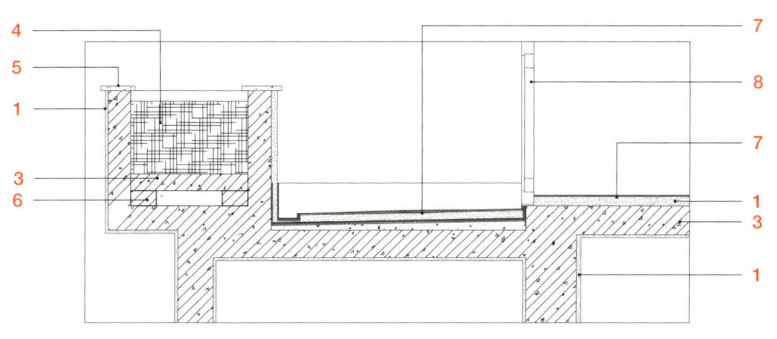

BALCONY SECTION

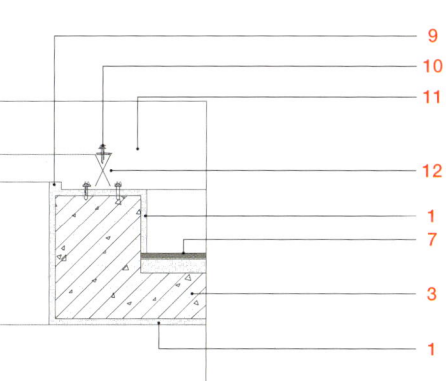

1 MORTOR
2 BLACK GRANITE STONE
3 CONCRETE
4 SOIL
5 BLACK NATURE STONE
6 BRICK
7 TILE FLOOR
8 DOOR
9 WEATHER STOP
10 SCREW
11 GLASS
12 STEEL BOX 50X100mm

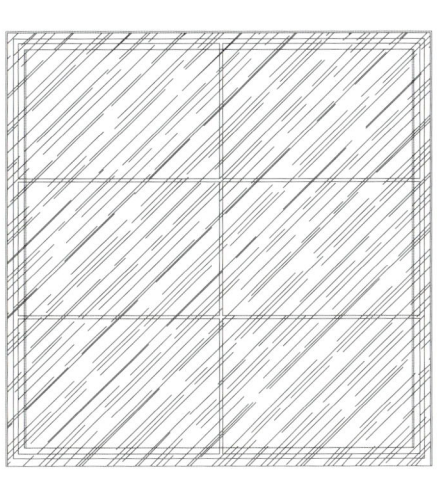

DETAIL OF GLASS ROOF

↑ 1st floor, vaccination center　↓ Stair

⌐ Void & the atrium
└ Void & the atrium

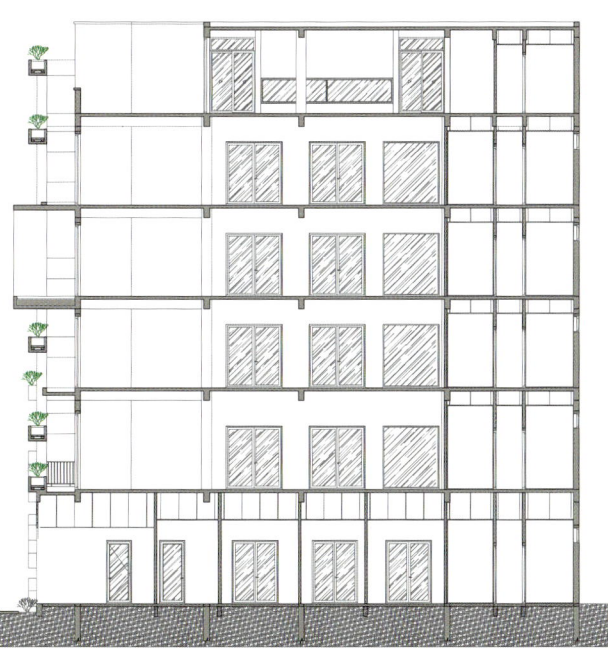
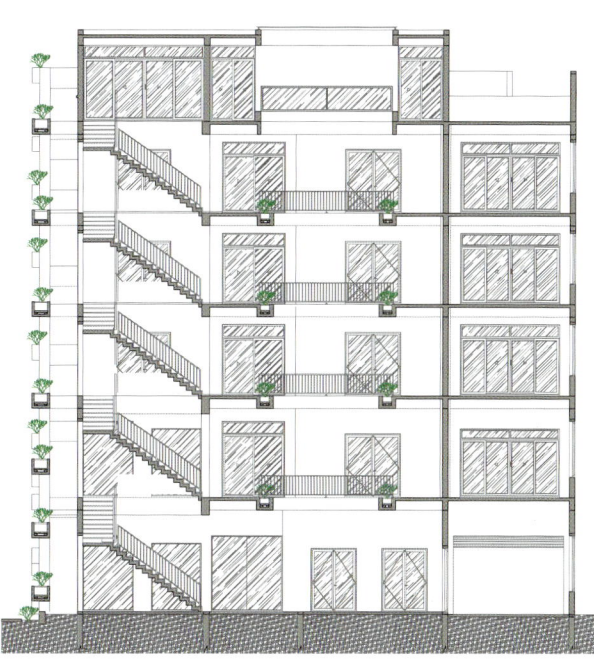

SECTION

↑ Void & atrium

1	RECEPTION	4	PHARMACY	7	PARKING SPACE	10	BALCONY
2	RELAX AREA	5	SECURITY ROOM	8	ROOM	11	COFFEE SHOP
3	CLINIC	6	TOILET	9	VOID	12	TERRACE

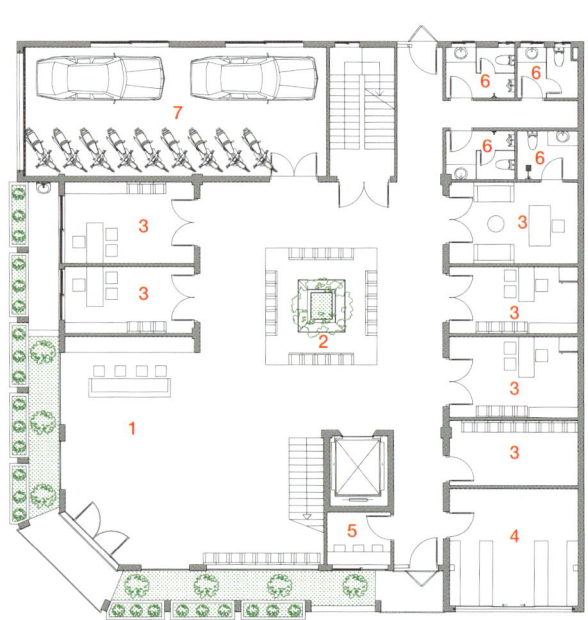

1ST FLOOR PLAN

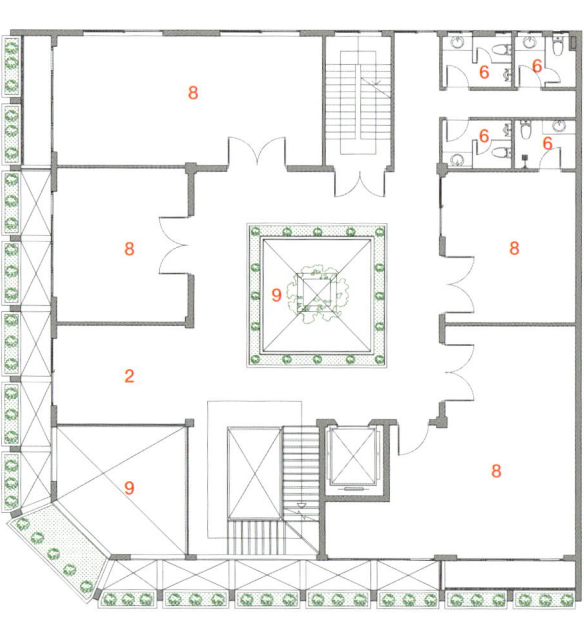

2ND FLOOR PLAN

← All floors can feel the greenery and natural light   → All floors can feel the greenery and natural light

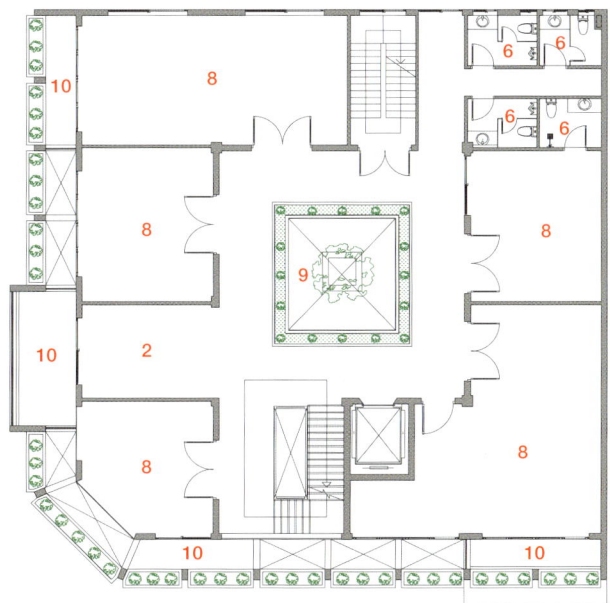

5TH FLOOR PLAN

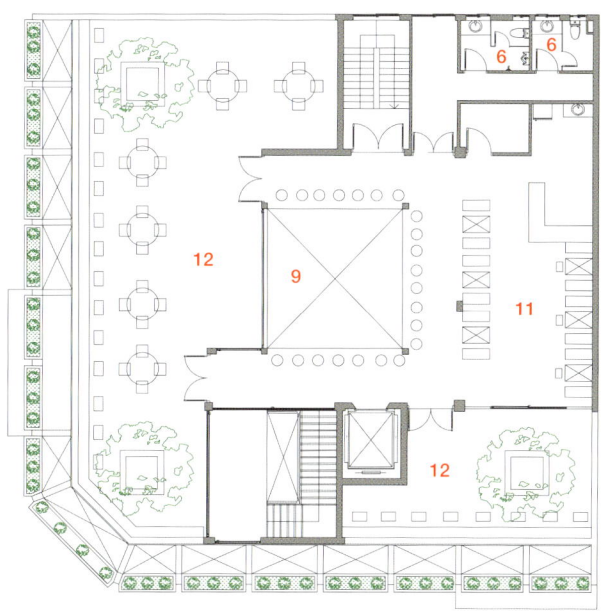

6TH FLOOR PLAN

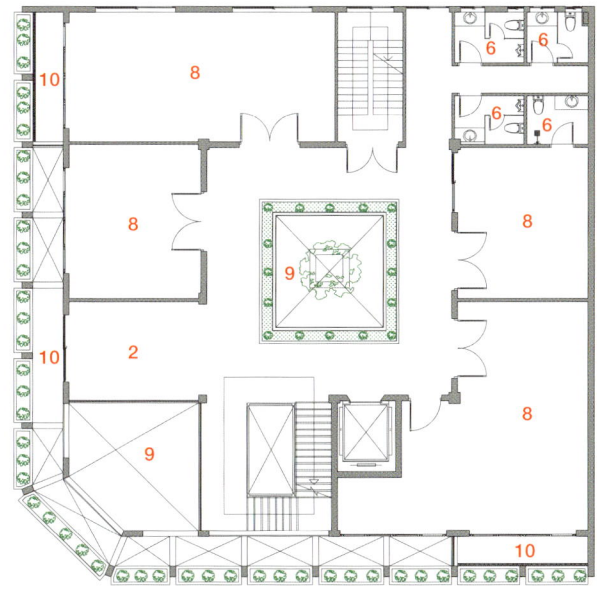

3RD FLOOR PLAN

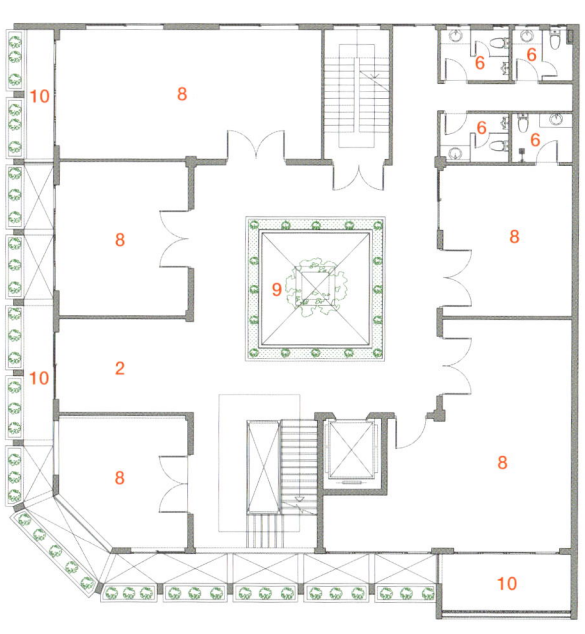

4TH FLOOR PLAN

유니텍스 사옥

UNITEX HEADQUARTER BUILDING

ARCHITECT : T2P ARCHITECTS OFFICE & P.O.T LAB / SHIKWAN YANG, TOMONORI MIURA, TATSUHITO ONO

T2P ARCHITECTS HAS RENOVATED UNITEX'S OLD BRANCH BUILDING to create the company's new headquarters in Tokyo, Japan. After the management of the 30-year-old-company changed, the Japanese architecture firm was presented with an opportunity to renovate the existing office, upgrade the space and represent the new face of the company. The X Office is UNITEX's new and refreshed headquarter space, featuring private offices, social spaces, and an open showroom for visitors to browse the company's LTO (data storage magnetic tape) products. The office features references to the company's identity throughout the space, most notably in the metallic mesh facade. The exterior is dominated by a large, draping mesh curtain that seals the offices behind it, while sliding mesh shutters on the ground floor, that are repeatedly marked with large Xs, enclose the product showroom.

An Enveloping Metallic Mesh Facade

For the renovation of the X Office, T2P Architects utilized the existing framework of the structure, while adding certain new elements to embody the company's spirit of learning from the past, and to upgrade their new image. The design team incorporates references to UNITEX's identity throughout the space in its design elements to achieve this.

An Enveloping Metallic Mesh Facade

In particular, the three-story building's previous tiles and glazed facade has been redesigned to hold a metal mesh exterior to represent the new face of the company. The upper two stories are sealed by a large, draping, metallic mesh curtain, while sliding security shutters marked with large Xs become a large part of UNITEX's identity and conceal the ground floor showroom. When fully closed at night, the shutters are semi-transparent and the building's entire front facade is enveloped in mesh, and when opened during the day they allow clear visibility into the interactive lobby area from the street front. Internally, an interactive community magnetic wall features repeated, small X shapes displayed in a uniform pattern in the step gallery, and stools with X shaped bases are placed around the room. On the staircase connecting the ground floor and basement, the balustrades consist of overlapping rods, forming a repeated X pattern.

Underground Parking Lot Converts To Meeting Spaces

Taking advantage of the site conditions and utilizing the space efficiently, T2P Architects converted the building's existing underground parking lot into storage spaces, a conference room, a multipurpose hall and meeting spaces. While the upper floors consist of private offices, the ground floor is transformed into a minimal, open showroom with social spaces where customers can freely browse products, and employees can meet and interact with each other.

Location Machida, Tokyo, Japan **Use** Office & Showroom **Site area** 274.60m² **Building area** 203.80m² **Gross floor area** 732.80m² **Completion** 2022 **Contractor** C.H.C. System Co., Ltd. **Photographer** Yukihide Nakano

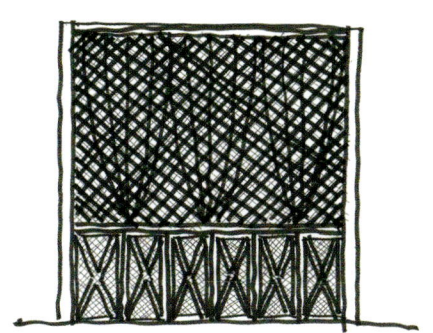

SKETCH

← Front facade

↑ Before & After of the facade

도쿄 마치다 시의 역사 깊은 '키느노미치(비단길)'에 위치한 창업 30년 된 LTO(데이터 보존 자기테이프)를 취급하는 기업이 2세대 경영을 기념하여 기존 지점 건물의 리모델링을 의뢰하였다. 기존 기계식 주차장이었던 1층과 지하층을 고객 전시와 직원 간 교류를 위한 공간으로 본사로서의 기능을 제안함과 동시에 금속 메쉬 파사드와 함께 새로운 회사의 얼굴이 되는 공간으로 디자인하였다.

기록 매체의 자기테이프를 다루는 본사로서 테이프의 곡선을 연상하는 드레이프 형태의 금속 메쉬로 기존의 타일 외벽을 덮고, 신구 요소가 오버랩되는 외관이 거리의 기억을 계승하도록 하였다. 금속 메쉬는 직사광선을 확산해 열부하를 줄이는 환경 장치이기도 하고, 흰색을 기조로 한 내부공간이 확산광을 안쪽으로 이끌도록 계획하였다. 회사 로고인 X를 모티브로 한 1층 방범 셔터는 내부 사용법에 따라 개폐함으로써 거리와 교감하도록 제안하였다.

내부는 기존 기계식 주차장의 큰 공간을 살렸고, 보이드를 통해 지하라는 폐쇄감을 느끼지 않는 일체적인 교류 공간으로 디자인하였다. 전시 부분은 목재 마감을 이용하여 친근감을 유도하였으며, 일부는 자석을 이용해 직원들이 자유롭게 교감할 수 있는 전시벽으로 계획하였다. 또한 오픈 키친 등 대화를 유발하는 다양한 공간적 장치가 맞물려 자연스러운 교류를 촉진하도록 하였다. 설비 면에서는, CO_2 센서에 의한 자동 환기와 일체형 제습·가습기에 의한 습도의 자동 제어로, 공조 부하를 경감해 에너지 절약화를 도모하였다.

각 부서의 디자인에 대해서는, 기업 아이덴티티나 제품을 상기하는 3개의 모티브를 전개하여 본사 빌딩 다운 개성 있는 공간 만들기에 유의하였다. 첫 번째는 회사명 UNITEX 로고에서 강조되는 'X'자 모티브로 파사드 방범 셔터나 1층 전시벽, 난간, 가구, 조명 등에 'X'자를 반복적으로 전개하였다. 두 번째는 자기테이프의 곡선을 모티브로 한 드레이프 형태의 금속 메쉬 파사드나 스테인리스로 만든 곡선 형상의 내외 간판과 각 실명의 사인은 직선적인 공간과 대비를 이루며 장소를 돋보이게 하였다. 세 번째는 감긴 자기테이프의 감긴적층 형태를 모티브로 하여 목재의 나이테를 연상시키는 가구와 전시 코너에 적층 합판을 이용하여, LTO라는 정보를 축적하는 기업 이미지를 표현하도록 하였다.

기존의 골격을 살리면서 기업을 브랜딩 하는 새로운 요소를 오버랩하여 회사가 내건 '온고 지신'의 정신을 공간에서 구현하도록 계획하였다. 그 공간 안에서 사내외의 교류가 지금까지 이상으로 활발하게 이루어져 LTO에 대한 정보를 발신해 나가는 거점이 되기를 기대한다.

SITE PLAN

↱ Partial view of exterior
↳ Partial view of exterior

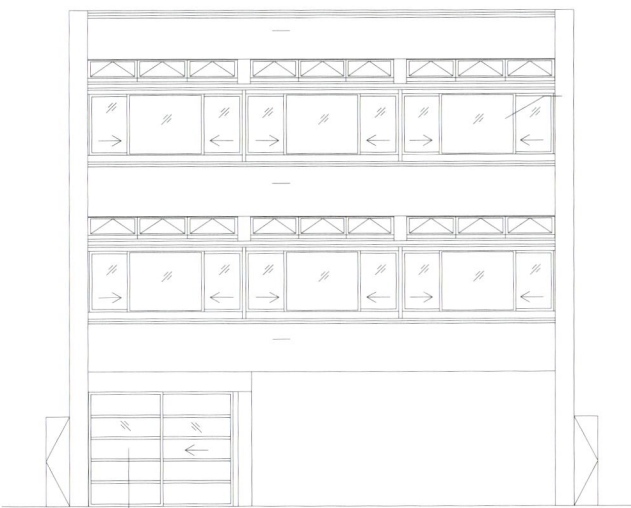

ELEVATION (BEFORE)

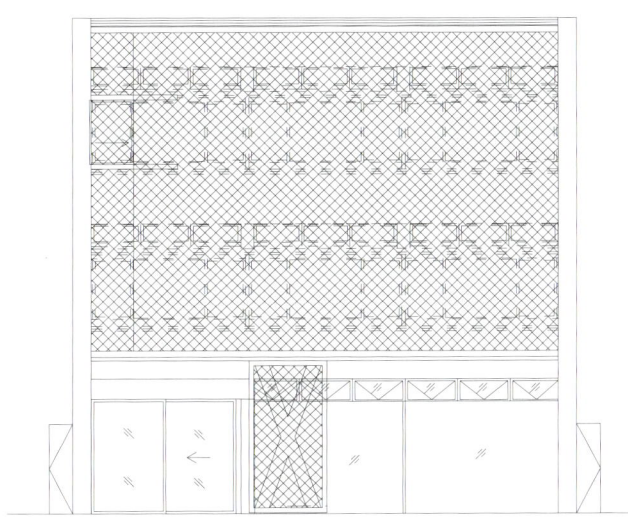

ELEVATION (AFTER - OPEN)

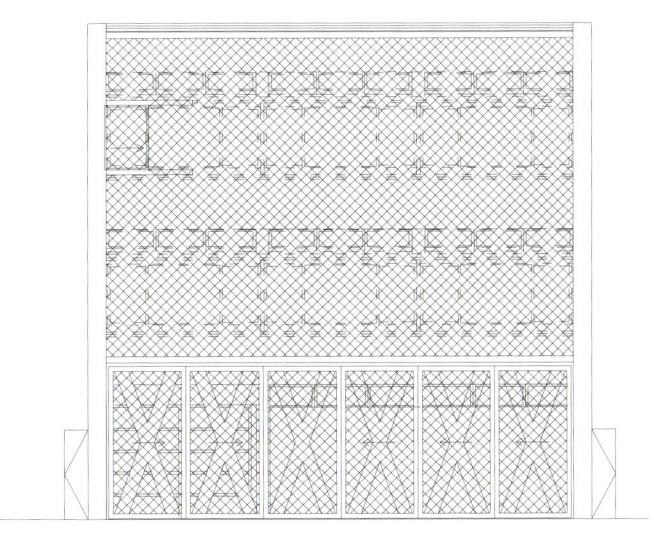

ELEVATION (AFTER - CLOSE)

↖ Lobby   ↙ X shutter   → X shutter   ↳ Variations of the X shutter

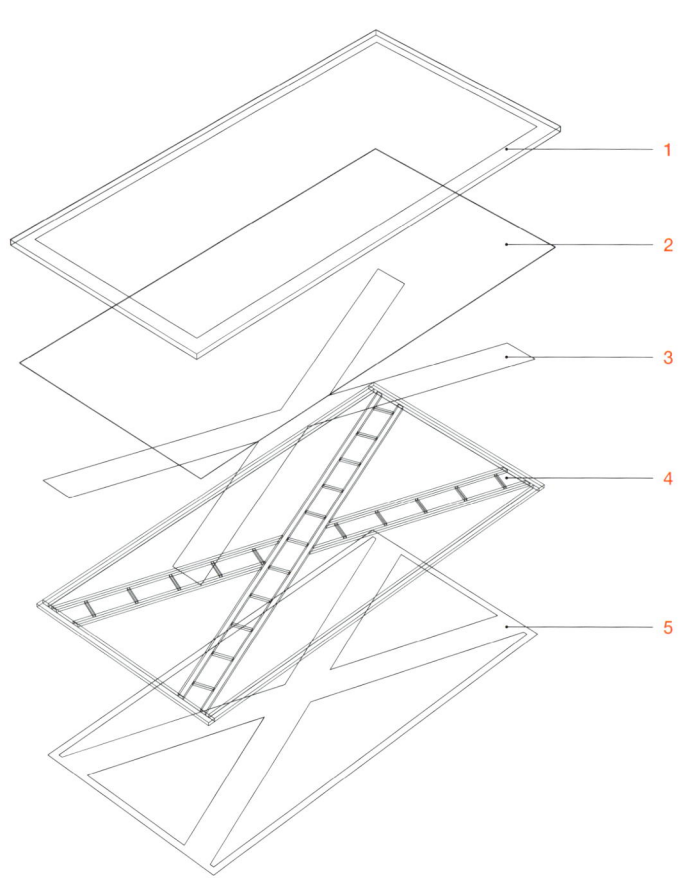

X SHUTTER STRUCTURE

↑ Entrance

1 COVER FRAME
2 STAINLESS STEEL MESH
3 WOODEN SHEET
4 PIPE STRUCTURE
5 WOODEN SHEET WITH COVER
6 T3 ALUMINUM COVER
7 OUTDOOR SG RAIL
8 RAIL CLIP
9 FACADE LIGHTING
10 300X150 H BEAM
11 X SHUTTER
12 M16 CHEMICAL ANCHOR
13 350X175 H BEAM
14 EXISTING TILE FACADE
15 WIRE CLIP IWC4
16 OUTDOOR RAIL

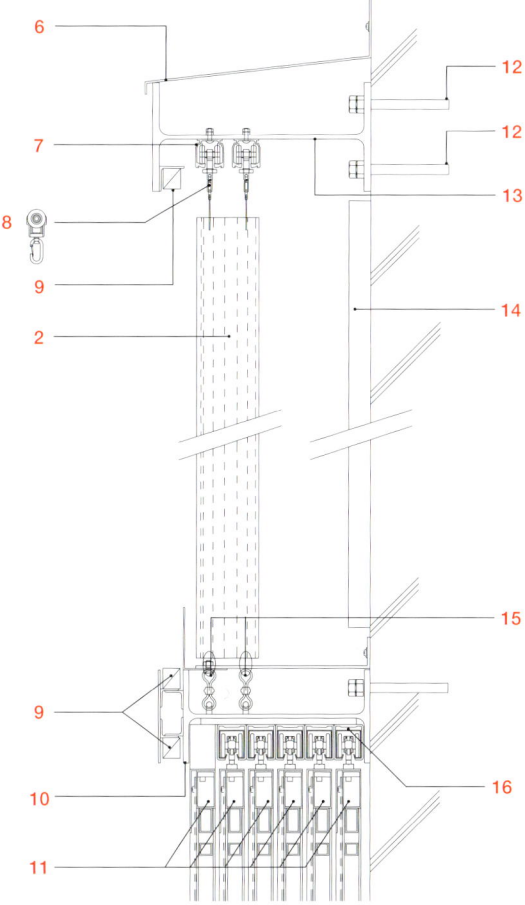

FACADE DETAIL

↑ Interior view of the 1st floor

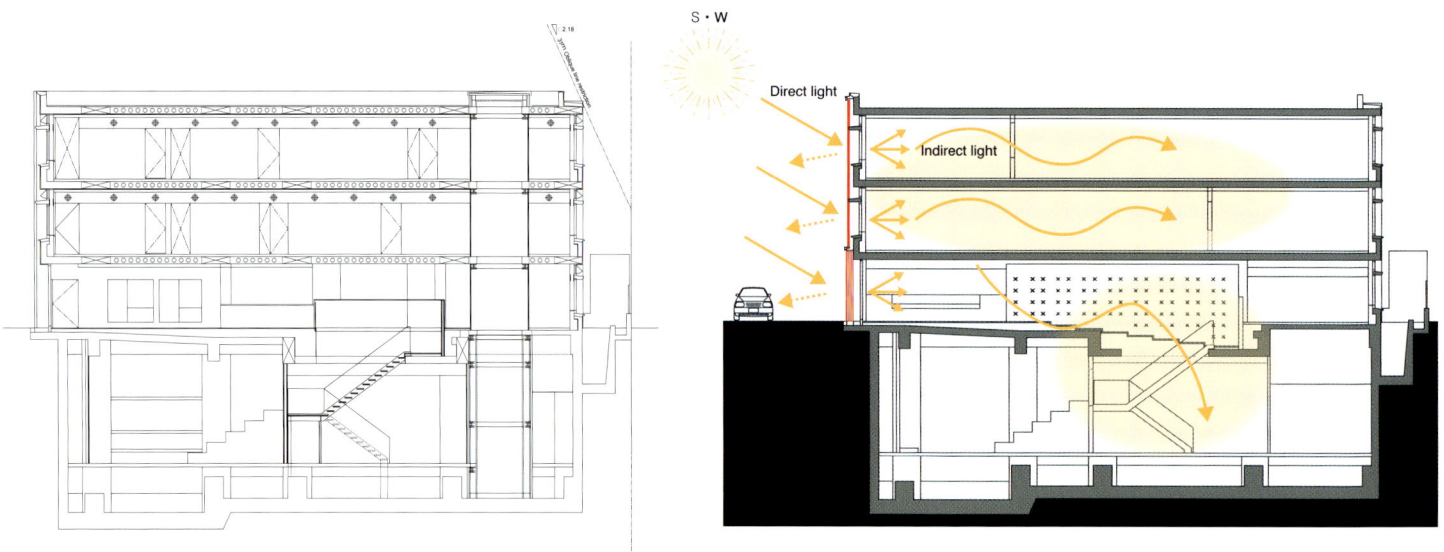

SECTION

CONCEPTUAL DIAGRAM OF ENERGY SAVING WITH DOUBLE FACADE

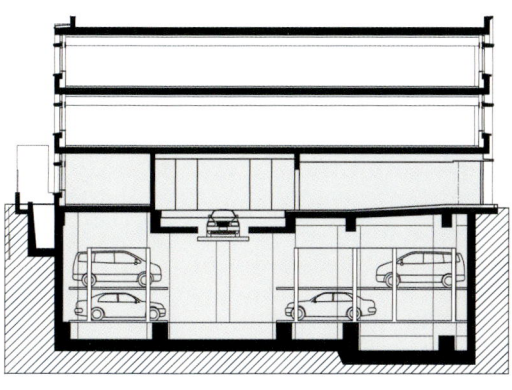
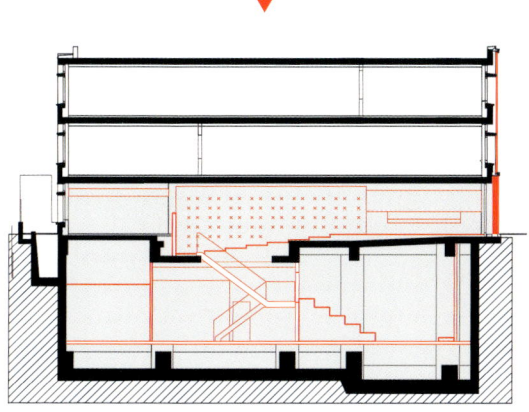
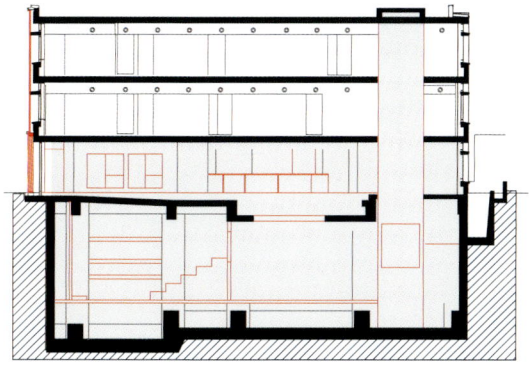

KEY SECTION

↑ Step gallery

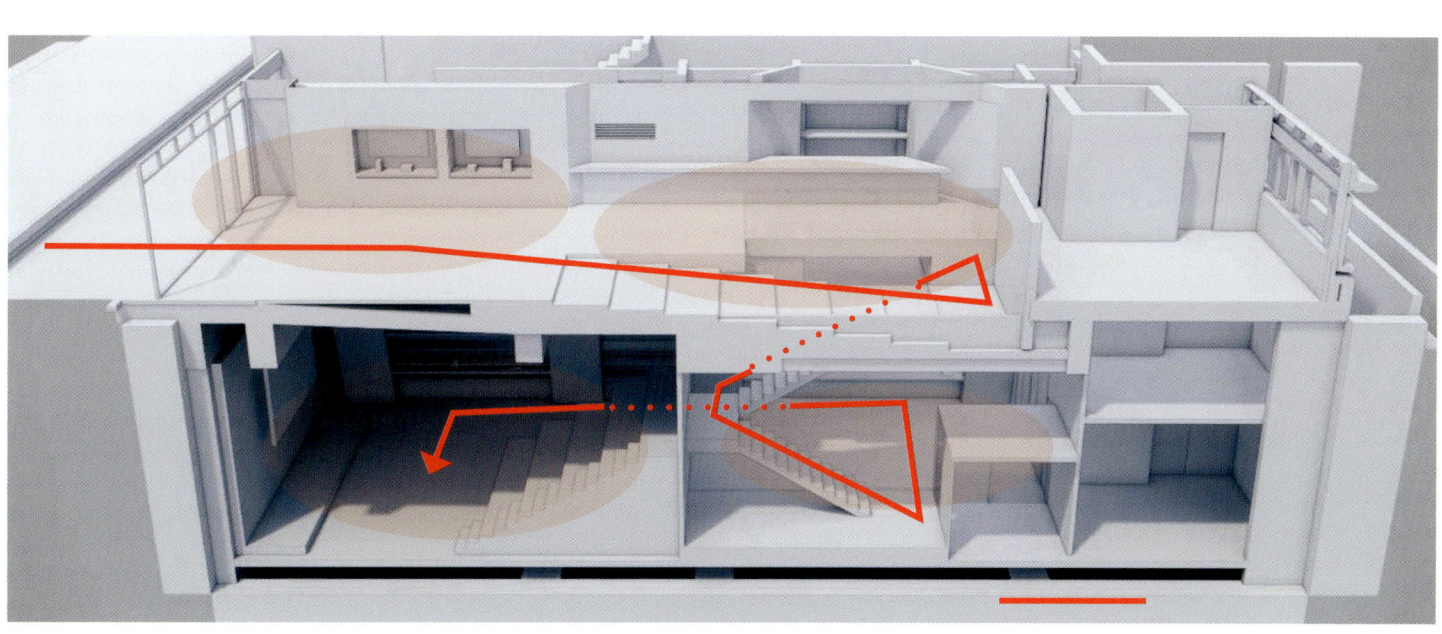

EXHIBITION CIRCULATION

↑ Step gallery   ↰ Lobby   ↲ Lobby   ↳ Lounge

1 RECEPTION	7 LOBBY	13 VENEER PLYWOOD	19 MAGNET
2 STORAGE	8 OPEN KITCHEN	14 VERTICAL FRAME	20 STEEL SHEET
3 SERVER ROOM	9 STEP GALLERY	15 T2 STEEL PLATE	21 MULTI-PURPOSE PANEL
4 SUB-KITCHEN	10 LOUNGE	16 AIR VENTILATION	
5 WC	11 INDIRECT LIGHTING	17 LGS FRAME	
6 EPS	12 CUSTOMIZED ACRYLIC MAGNETIC X PIN	18 X-SHAPED ACRYLIC, SUPPORT PIN	

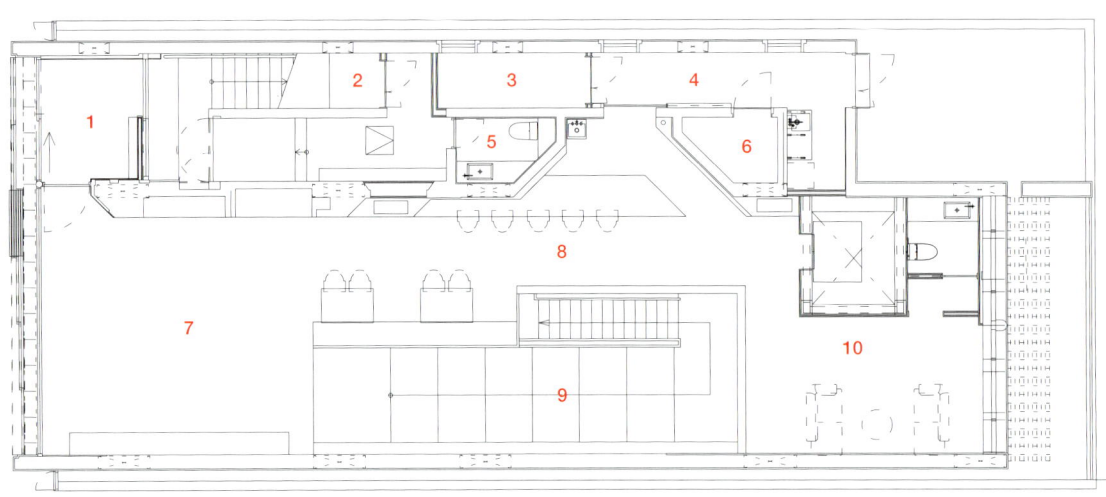

1ST FLOOR PLAN

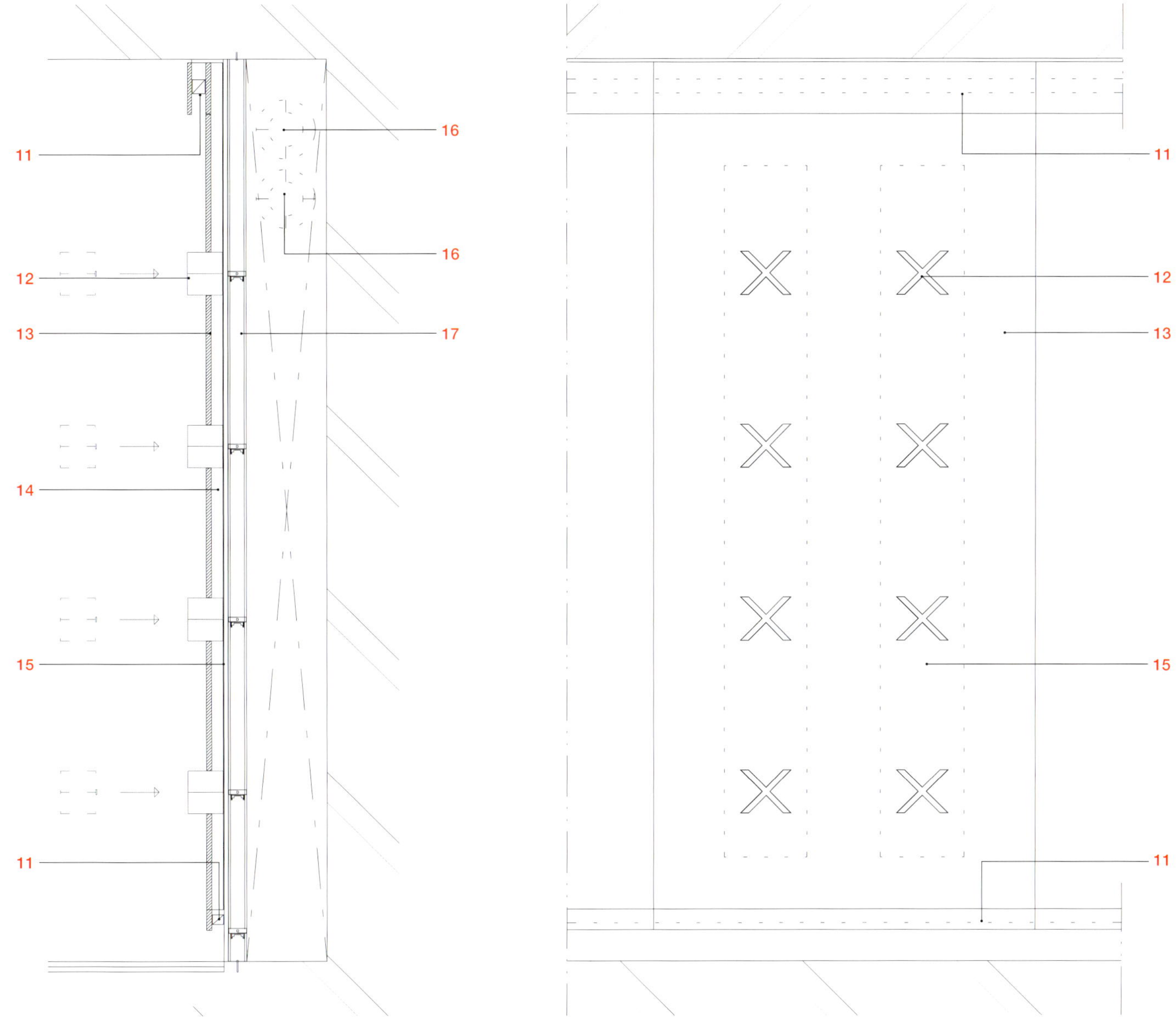

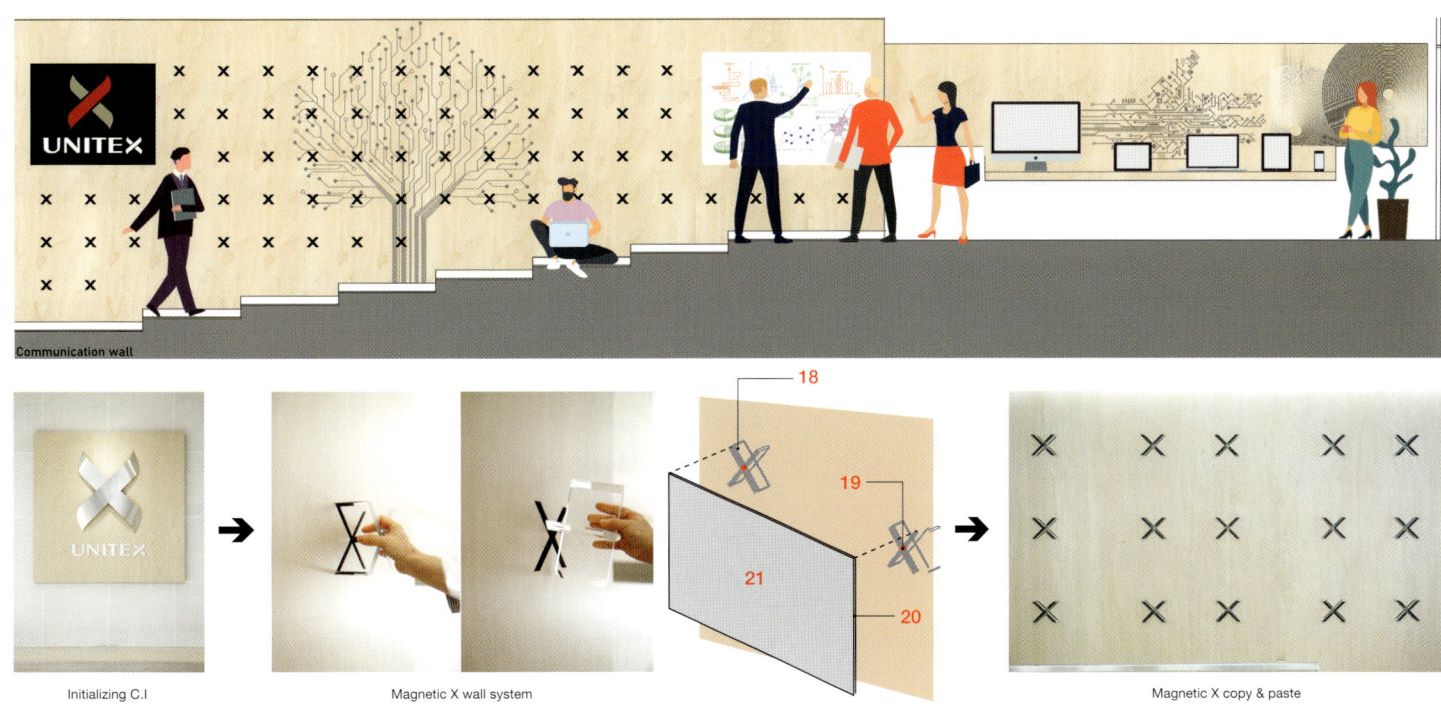

DETAIL OF EXHIBITION WALL

Initializing C.I

Magnetic X wall system

Magnetic X copy & paste

COMMUNICATION DIAGRAM WALL

↑ Corner view

| 1 | MULTI-PURPOSE HALL | 3 | STORAGE | 5 | LIBRARY LOUNGE |
| 2 | LOUNGE | 4 | MEETING ROOM | 6 | CONFERENCE ROOM |

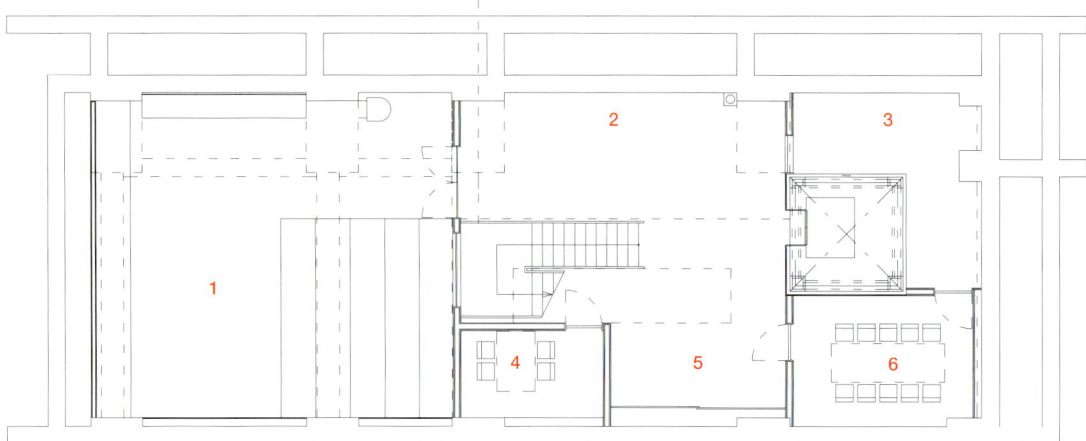

BASEMENT FLOOR PLAN

← Kitchen & dining room   → Ground floor stair   → Ground floor stair

음악의 빌딩
BUILDING OF MUSIC

ARCHITECT : AISAKA ARCHITECTS' ATELIER / KENSUKE AISAKA

THIS COMMERCIAL COMPLEX FOR A MUSIC COMPANY with a long history includes music classrooms, instrument sales and storage spaces, and the company's main offices. Located in the Kanda-Ochanomizu area of Tokyo, which is known for its many music stores, the lot is small and irregularly shaped. The client requested a building that efficiently combined a variety of functions, including rooms for lessons in piano, violin, and other instruments; a hall for recitals; a professional recording equipment shop; storage for wholesale instrument sales; and offices to manage these various aspects of the businesses.

The requirements related to soundproofing, views, daylighting, shade, and usage time varied for each floor, and the likelihood that these requirements would change in the future was high. We therefore decided to use a double skin that maintains a unified exterior appearance regardless of how rooms are being used or whether windows are open. We selected a mesh membrane that sways at high altitudes, can easily be rounded in all directions to allow for optimal sky-view ratios, and is porous to light and air. Made of polyester fibers coated with PVC resin, the fireproof exterior mesh membrane blocks visibility from the outside while allowing those inside to see out. A solar reflectance rate of 33 percent lowers the air conditioning load by 28 percent. The membrane can be bent three-dimensionally, and because it is light and perforated, there is little structural stress from either the weight of the material itself or wind pressure. Because there is no danger related to the material falling to the ground, it is easy to replace and maintain and can be reused. The COVID-19 pandemic has increased the appeal of facades such as this that enhance natural ventilation while maintaining privacy.

The form of the building takes into account the irregular lot shape, codes related to height and setback, and necessary elements such as emergency stairs, balconies, plumbing, and wiring as well as the need to express the company's identity. We incorporated shapes inspired by various musical instruments, adjusted to maximize capacity. The mesh-covered facade evokes a speaker sending music into the neighborhood, transforming the building into a giant sign advertising the company inside.

Location Chiyoda-ku, Tokyo **Use** Office & Acamedy **Site area** 351.67m² **Building area** 220.90m² **Gross floor area** 1,667.98m² **Completion** 2022 **Structure** Ohno Japan **Equipment** Zo Consulting Engineers **Lighting Design** Izumi Okayasu **Lighting Design Sign Design** Hiromura Design Office **Photographer** Shigeo Ogawa, Kensuke Aisaka

← The framework of the mesh facade   → Forms reference various instruments

↑ Day and Night views   → Mesh membrane like a speaker

오랜 역사를 가진 음악 회사의 상가에는 본사와 함께 음악 학원과 악기 매장, 보관실이 함께 자리하고 있다. 부지는 많은 음악 매장으로 유명한 도쿄의 칸다와 오차노미즈 지역에 위치해 있으며, 크기가 작고 불규칙한 형태를 이룬다. 건물 소유주는 이 건물에 피아노와 바이올린 및 기타 악기 레슨실은 물론 음악 공연장, 전문 녹음 장비 상점, 도매상 악기 보관실 및 행정 사무실 등 다양한 기능을 갖춘 설계를 요청해왔다.

층별로 방음과 전망, 채광, 자외선 차단 및 사용 시간에 대한 요구사항이 다양했으며, 이는 향후 변경될 가능성이 농후했다. 이에 방의 용도나 창문의 개폐 여부와 관계없이 통일된 외관을 유지하기 위해 이중으로 설계하기로 했다. 그리고 높은 고도에 적합한 유연함을 갖추고 모든 방향으로 곡선을 이루어 최적의 전망을 제공하며, 다공성 구조로 채광 및 환기를 향상시켜주는 메쉬 멤브레인을 선택했다.

PVC 수지로 코팅된 폴리에스터 섬유로 제작된 내화성 메쉬 멤브레인으로 외부를 마감해 안에서는 밖을 볼 수 있지만 밖에서는 내부를 볼 수 없게 차단해준다. 태양광 반사율이 33%로, 에어컨 부하를 28%까지 낮췄다. 3차원의 곡선을 이루는 멤브레인은 가벼운 다공성 구조로 소재 자체의 무게나 풍압에 의한 구조 응력이 거의 발생하지 않는다. 자재가 땅에 떨어질 위험이 없어 교체 및 보수가 용이하고 재사용도 가능하다. 코로나19 팬데믹으로 인해 프라이버시를 유지하면서도 자연적인 환기를 향상시키는 파사드에 대한 요구가 높아졌다.

빌딩의 모양을 디자인할 때 불규칙한 형태의 부지와 높이 및 건축 제한선 관련 규정, 비상계단, 발코니, 배관 및 배선 등 필수적인 요소는 물론 기업의 아이덴티티를 고려했다. 그래서 다양한 악기에서 영감을 얻어 최적의 기능성을 갖춘 통합적 형태의 건축물을 탄생시켰다. 메쉬로 덮인 파사드는 주변으로 음악 소리를 전달하는 스피커를 연상시키며 건물 자체가 회사의 정체성을 드러내는 하나의 거대한 간판 역할을 한다.

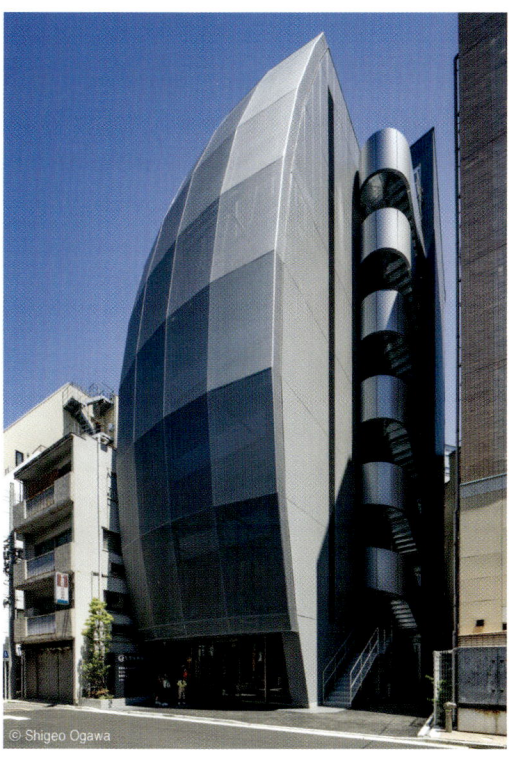

1. COPING : GALVANIZED STEEL PLATE
2. MESH MEMBRANE
3. CEILING : INSULATION MAT
4. FLOOR : TILE CARPET
5. FLOOR : VINYL COMPOSITION TILE
6. DECK PLATE
7. FLOOR : TILE
8. CEILING : ROCK WOOL BOARD
9. FLOOR : WOODEN FLOORING
10. MAIN RECITAL HALL
11. ENTRANCE HALL
12. LECTURER WAITING ROOM
13. LOBBY
14. LESSON ROOM
15. STORAGE
16. RECORDING EQUIPMENT SHOP
17. CAFE CORNER
18. OFFICE

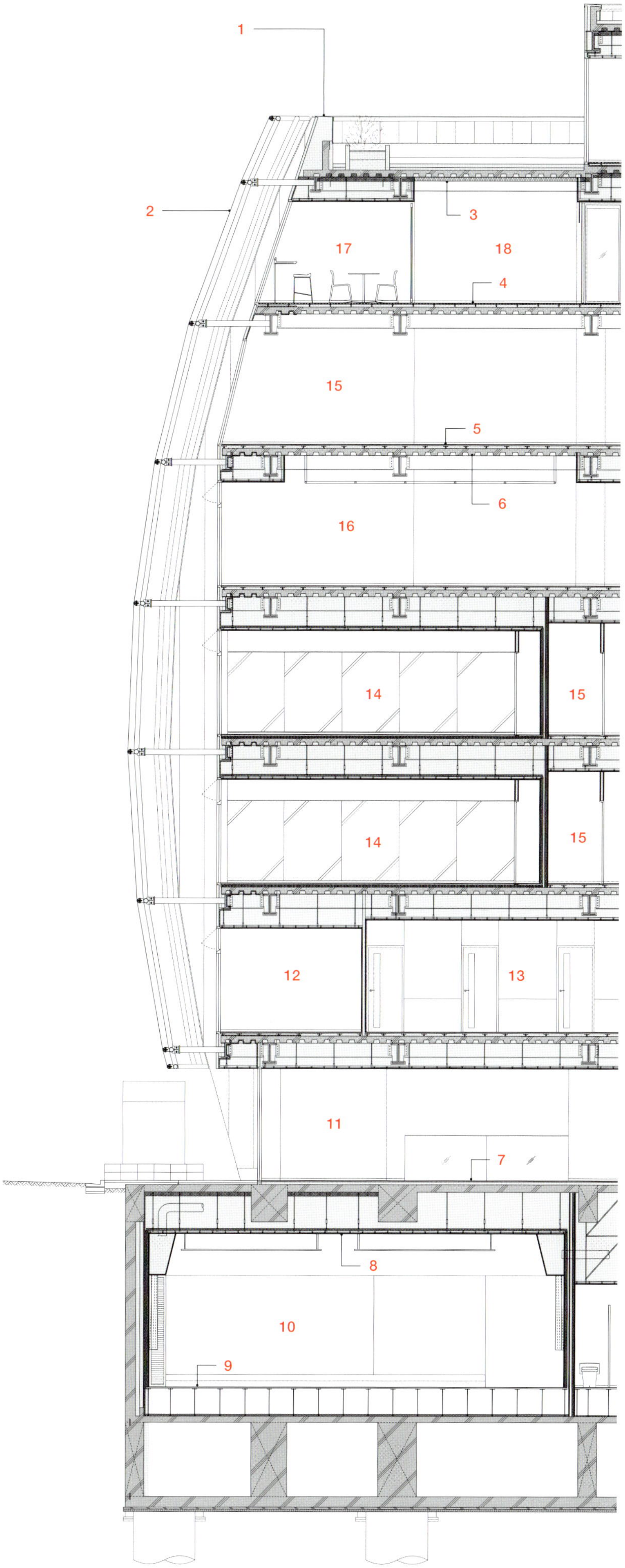

SECTION

↑ Storage for wholesale   ↓ Facade frame

1. T0.95 MESH MEMBRANE, POLYESTER FIBERS+PVC RESIN
2. HORIZONTAL FRAME
3. VENTICAL FRAME
4. ALUMINUM
5. CANTILEVER
6. EXTRUDED CEMENT PANEL, FLUORORESIN PANT
7. T16 PLATE
8. 2-M20(F10T)
9. T9 GALVANIZED PLATE

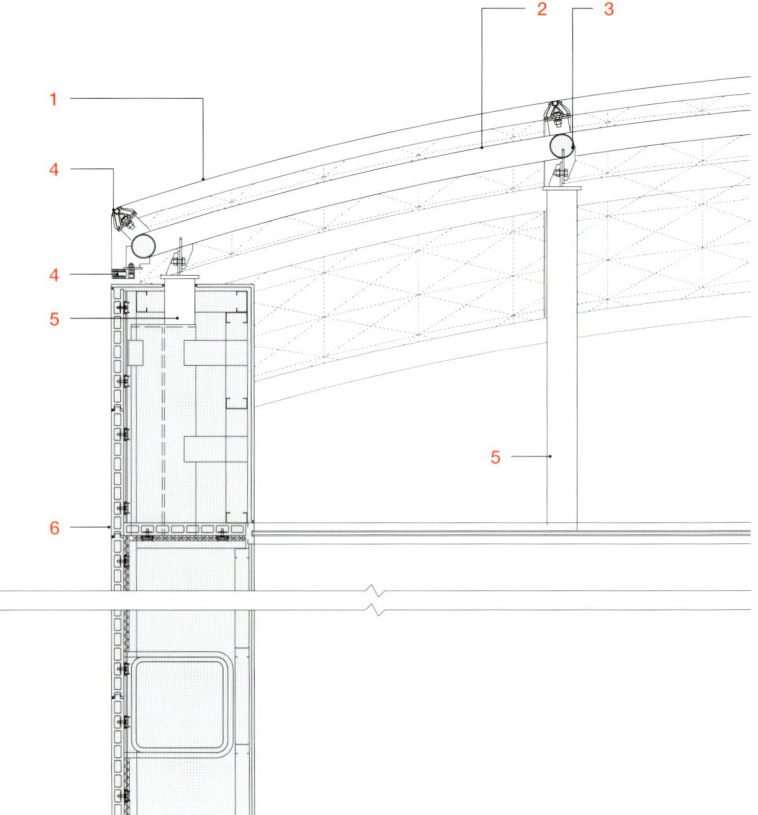
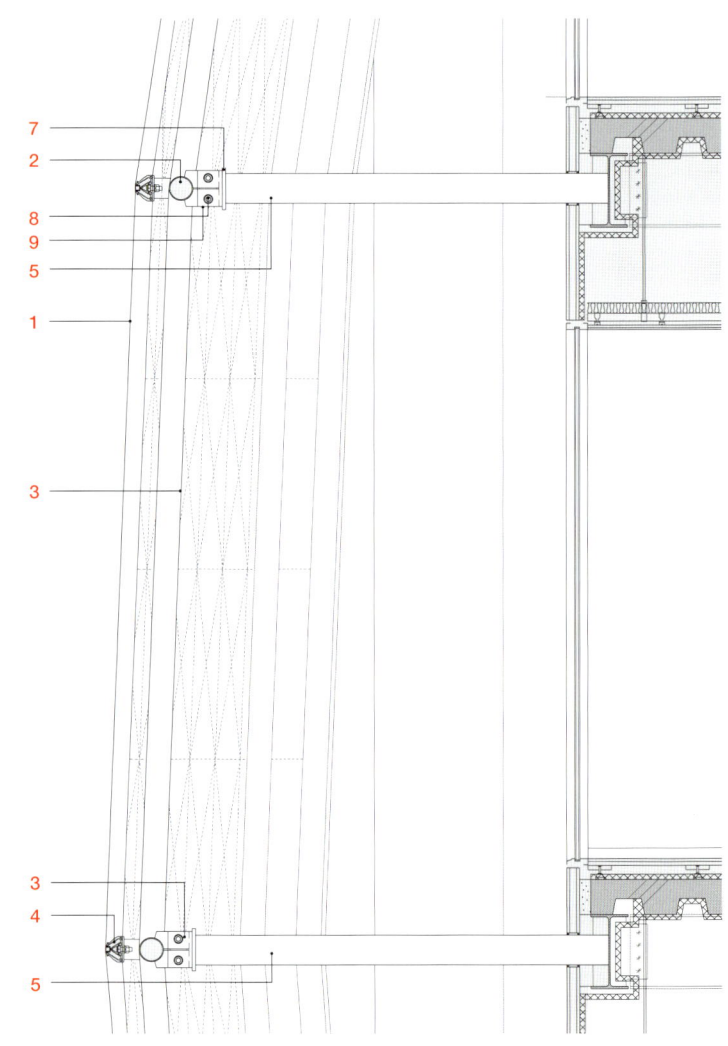

SECTION DETAIL

↑ Cafe corner on the 7th floor

1 STAGE
2 RECITAL HALL
3 BACKSTAGE
4 DRESSING ROOM
5 DRUMS LESSON ROOM
6 ENTRANCE HALL
7 RECEPTION
8 SUB ENTRANCE
9 OFFICE
10 STORAGE
11 GARBAGE ROOM
12 STUDIO
13 LESSON ROOM

2ND FLOOR PLAN

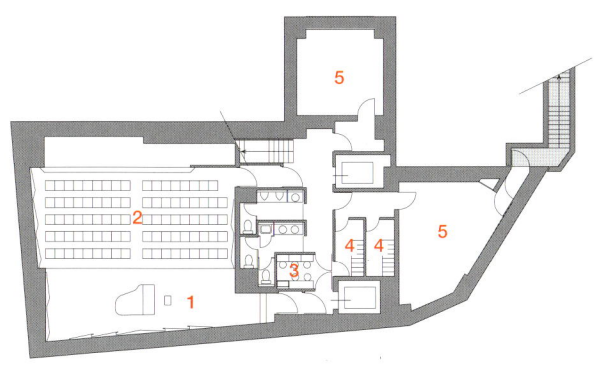

BASEMENT FLOOR PLAN

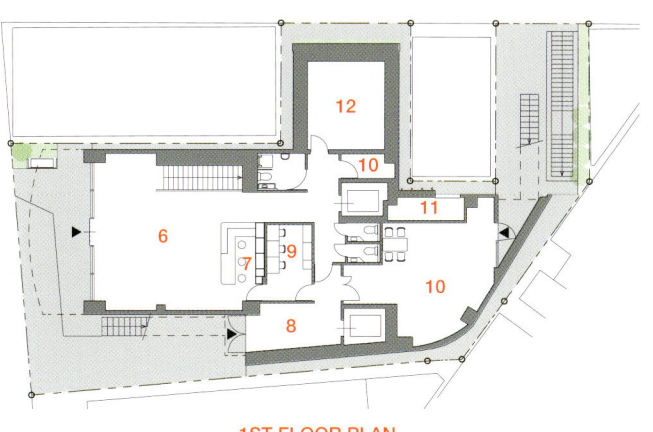

1ST FLOOR PLAN

← Entrance hall → Main recital hall

1 LESSON ROOM
2 DRESSING ROOM
3 RECORDING EQUIPMENT SHOP
4 ELECTRIC ROOM
5 RECEPTION
6 DEMONSTRATION STUDIO
7 STORAGE
8 OFFICE
9 CONFERENCE ROOM
10 CAFE CORNER
11 TERRACE
12 MULTIPURPOSE ROOM
13 EXECTIVE ROOM

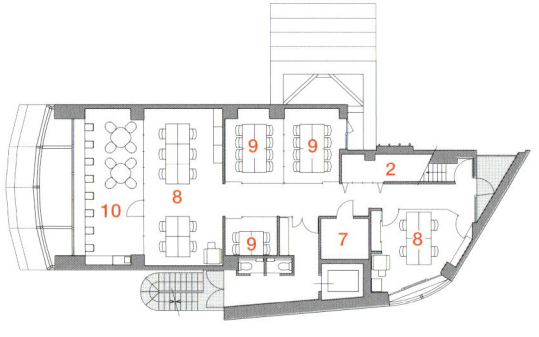

7TH FLOOR PLAN

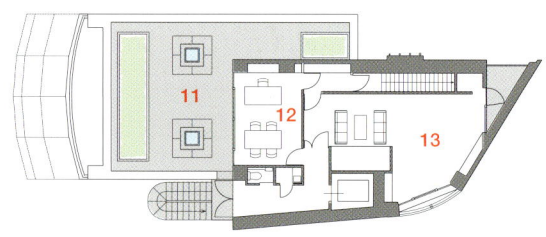

ROOF FLOOR PLAN

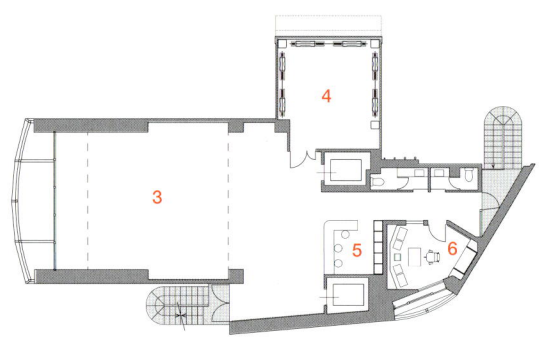

5TH FLOOR PLAN

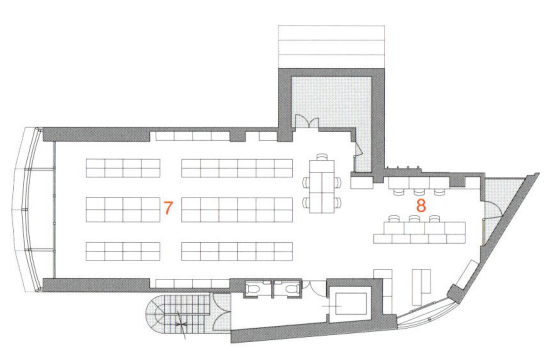

6TH FLOOR PLAN

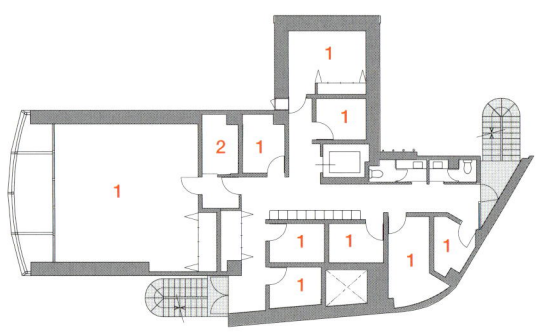

3RD FLOOR PLAN

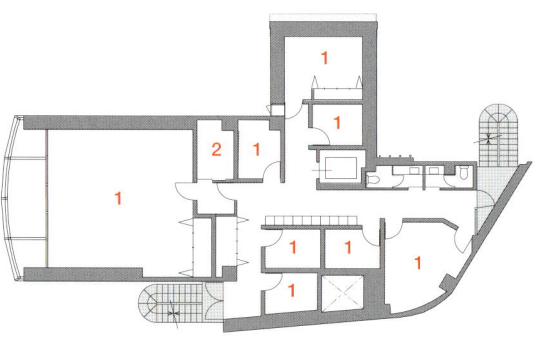

4TH FLOOR PLAN

↱ Terrace on the 8th floor
→ Office
↳ Recording equipment shop

국립미술공예디자인센터

NATIONAL CENTRE FOR ART, CRAFTS & DESIGN

ARCHITECT : RAMOS CASTELLANO ARQUITECTOS / ELOISA RAMOS, MORENO CASTELLANO

THE WORKS OF THE CNAD NATIONAL CENTRE for Art, Crafts and Design in Mindelo, port city of Cape Verde, on the island of Sao Vicente, had to be handcrafted, and not only for the name and the function it was to perform, but above all because entrusting the manual works locally would have meant distributing the public funding, resulted from a great government effort, among the artisans of the isle. Thus fulfilling, all the more so since it is a public center, one of the primary functions of architecture, namely that of being useful to the community. Every design choice, from materials to finishes, has been aimed at seeking to achieve this goal.

In this part of the world the industrial revolution, meant as the transition from manual labor to mechanization, is slowly starting now. Here there still exist, and they are the majority, craftsmen who daily build artifacts for the people. Every constructive element of the CNAD is the result of this type of craftsmanship, the entire building is a great artisan work.

The construction and assembly work was carried out entirely by hand with the aid of machines which can be assimilated to simple working tools. In the metal structure, on which the prefabricated elements were mounted, welded and assembled by the carpenters, the architects have riveted 2,532 drumheads, recycled from old metal drums, which have been sandblasted and manually painted one by one, to form a vibrant, colorful ventilated facade.

The culture of recycling is an integral part of the life of these islands, which know how to treasure every resource: here empty drums are never wasted, they are opened and transformed into metal sheets to cover the houses, used as formwork for the concrete casting, reworked to obtain pots and knives. The idea of using them as a base material for the restoration sounded perfect. In the new building – combined with the restored colonial house built as the residence of Senator Augusto Pereira Vera-Cruz, then transformed into a high school and, subsequently, into the headquarters of the historic radio Barlavento where Cesaria Evora recorded her first songs – masons and carpenters have worked together to achieve the result of a great work of local craftsmanship, in which imperfection is tangible, and it is seen as an added value, not a defect, because it conveys a sense of manufacturing, of uniqueness.

The colors hide a musical score. Each colors is a musical note. They wanted to use the perceptual phenomenon of synesthesia so a Caboverdian composer and multi-instrumentist, Vasco Martins, was invited to participate and wrote the music behind the colors. Paying homage to the musical traditions of the islands and transmitting a visual musical joy to the square.

For the CNAD project they have taken inspiration from the people that live in the city of Mindelo, from the Cabo Verdian manner to find solution to difficult problems, and from Chirsto and Jeanne Claude barrel sculptures. They are constantly in the search of harmony. The same joy that people receive through the building that we imagine, came back to us in multiple ways.

Location Mindelo, Sao Vicente Island, Cabo Verde **Use** Culture & Public **Site area** 1,265m² **Building area** 1,450m² **Gross floor area** 1,450m² **Completion** 2022 **Project manager** Eloisa Ramos, Moreno Castellano **Design team** Zico Lopes, Bruno Kenny, Edoardo Meneghin, Marvin Delgado, Danil Silva, Marco dos Anjos **Contractor** SGL - Sociedade de Construcoes SA **Photographer** Sergio Pirrone

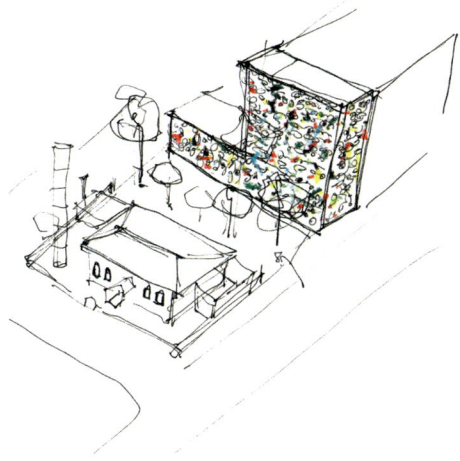

SKETCH

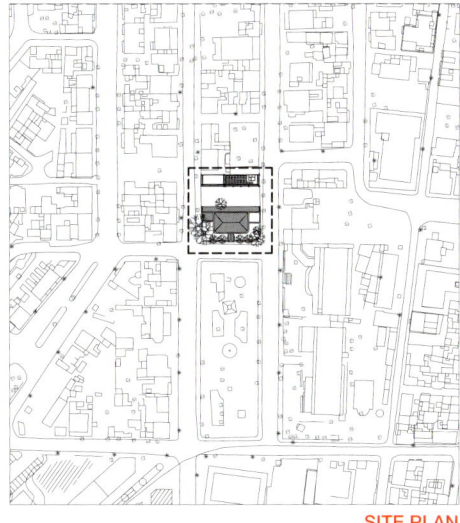

SITE PLAN

← Panoramic view

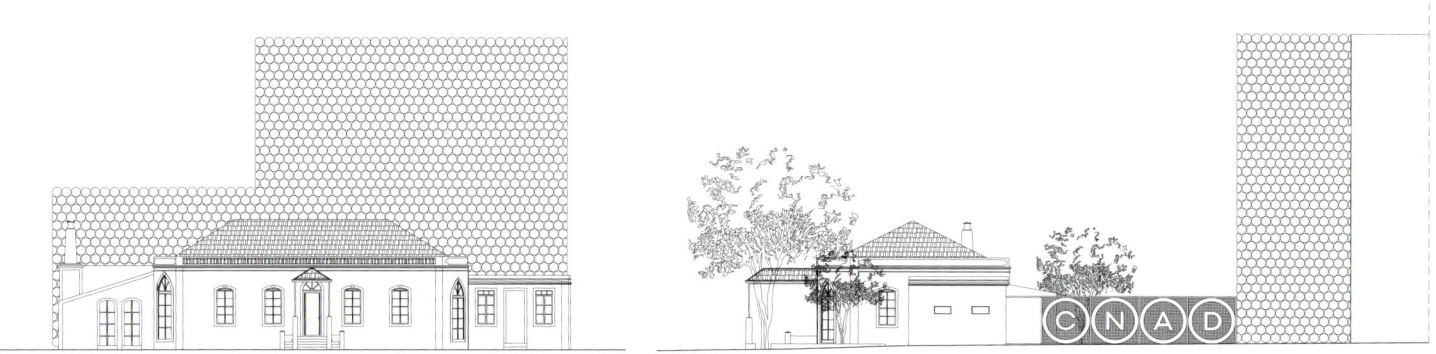

↑ Panoramic view

FRONT ELEVATION

RIGHT ELEVATION

카보베르데 제도에 위치한 상비센트 섬의 항구 도시 민델루에 세워질 국립 예술, 공예, 디자인 센터(CNAD) 프로젝트는 이름뿐 아니라 건물의 특성과 더불어 무엇보다도 정부의 거대한 공적 자금을 현지 장인들에게 분배하기 위해서도 수작업으로 진행해야 했다. 본 건물은 공공 센터이므로 건축의 주요 기능인 커뮤니티에 유용함을 주는 전략이기도 하다. 따라서 재료를 선정하고 마감을 할 때까지 건축의 초점은 이 목표를 달성하는 데 두었다.

이 지역에서는 육체노동에서 기계화로의 전환을 의미하는 산업 혁명이 천천히 시작되고 있었다. 동시에 사람들을 위해 매일 공예품을 만드는 장인들 역시 여전히 주류로 활동하고 있다. CNAD를 구성하는 모든 요소는 장인의 손길을 거친 것으로, 빌딩 전체가 하나의 거대한 작품 이다.

건설 및 조립 과정은 전적으로 수작업을 통해 진행되었으나, 일부 단순 작업에 있어서는 기계의 도움을 받았다. 목수가 사전 성형된 부품을 용접을 통해 조립한 금속 구조물의 경우, 2,532개의 오래된 금속 드럼통의 뚜껑을 리벳으로 고정해 다채롭고 생동감 넘치며 통기성이 뛰어난 파사드를 탄생시켰다.

자연을 소중하게 여기는 방법을 알고 있는 이 섬에서는 재활용 문화가 일상에 녹아들어 있다. 빈 드럼통을 버리지 않고 분리해 집을 덮는 금속판으로 사용하거나, 콘크리트 주조를 위한 거푸집으로 활용하기도 하고 냄비나 칼의 재료로 쓰기도 한다. 이번 재건축 프로젝트에서 이것을 기본 재료로 쓰는 것은 완벽한 아이디어였다. 식민지 주택을 아우구스토 페레이라 베라-크루즈 상원 의원의 집으로 개조했다가 고등학교가 된 후 마침내 세자리아 에보라가 첫 노래를 녹음한 역사적인 라디오 방송국 발라벤토의 본사가 된 경우와 마찬가지로 새로운 프로젝트에 투입된 현지 목수, 장인들과의 협업을 통해 위대한 작품이 완성될 수 있었다. 눈에 띄는 불완전한 부분은 결함이 아니라 독특한 감각의 가치가 빛나는 장인정신이다.

다채로운 색은 악보를 숨기는 대신 그 자체로 하나의 음표가 된다. 이 콘셉트가 공감각적으로 전달될 수 있도록 카보베르데 출신의 작곡가이자 다양한 악기 연주자인 바스코 마틴을 초청해 강렬한 색채 뒤에 음악적 요소를 넣었다. 섬 전통 음악에 대한 경의를 표하는 동시에 다채로운 색으로 광장에 음악이 울려 퍼지게 하였다. CNAD 프로젝트를 완성하기까지 민델루 시 주민들의 일상과 어려움을 해결하기 위한 카보베르데만의 방식, 그리고 드럼통을 활용한 크리스토와 장 클로드의 조각품에서 영감을 얻었다. 그들은 끊임없이 조화로운 삶을 가꿔간다. 그들이 상상한 건축물을 통해 누리는 기쁨을 건축가도 충분히 느낄 수 있었다.

↱ Corner view
→ Bird's eyes view
↳ View from south

↰ Street view
↱ Remodeled house
↓ Upper part of the facade

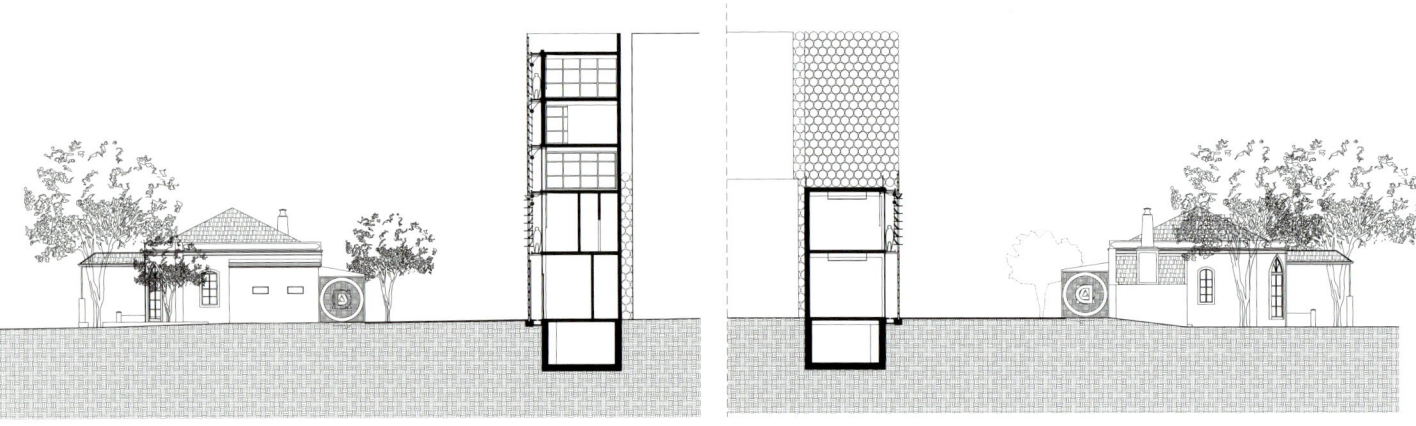

SECTION

● RED	· RAL 3020	· note Do
● ORANGE	· RAL 2004	· note Re
● GREEN	· RAL 6001	· note Mi
● DARK BLUE	· RAL 5017	· note Fa
● DARK YELLOW	· RAL 1028	· note Sol
● BLUE	· RAL 5015	· note La
● VIOLET	· RAL 4001	· note Si
● YELLOW	· RAL 1023	· notes together
● BLACK	· RAL 9005	· chord
○ WHITE	· RAL 9003	· pause
● DARK RED	· RAL 3003	· drone
● CIAN	· B621	· sequence
● BROWN	· RAL 8004	· rhythm
● OLIVE GREEN	· RAL 1027	· bass

LIST OF COLORS & NOTES

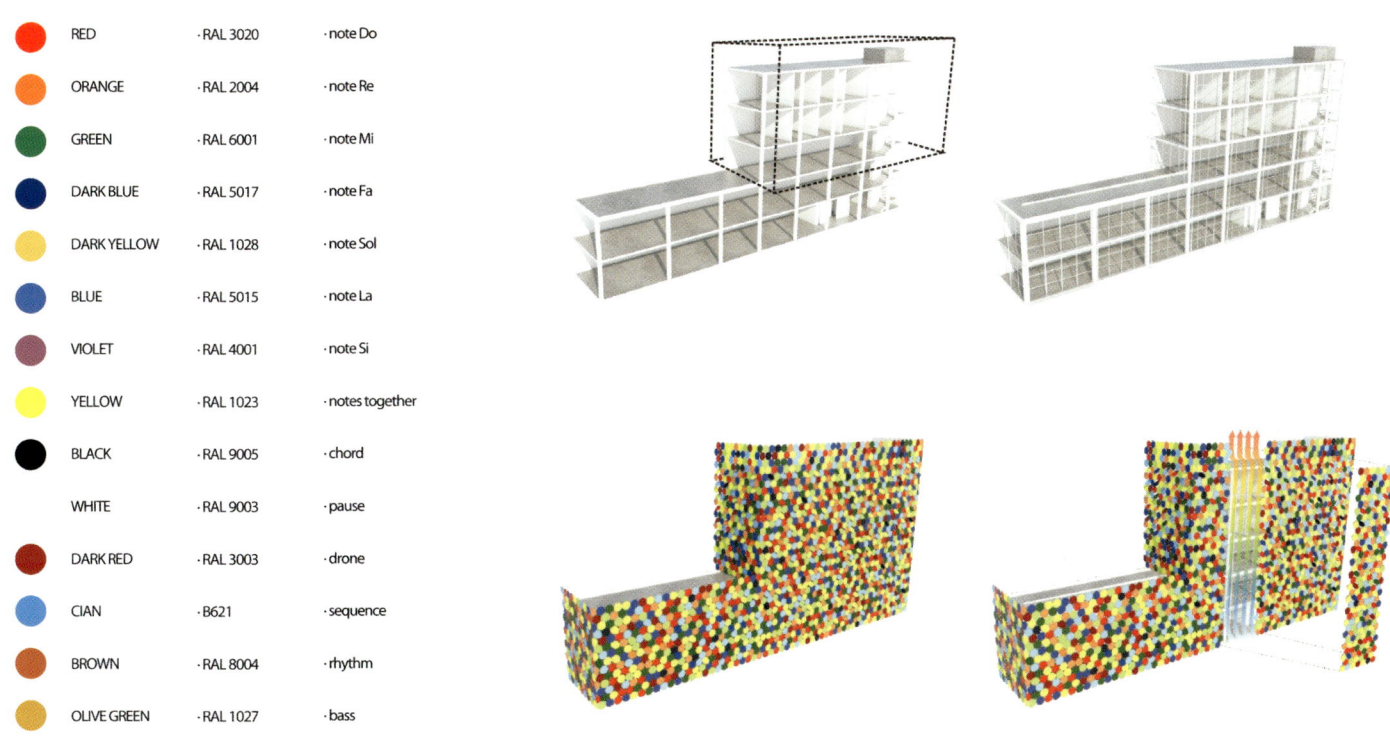

FACADE VENTILATION

>> RECYCLING CAPS PROCESS

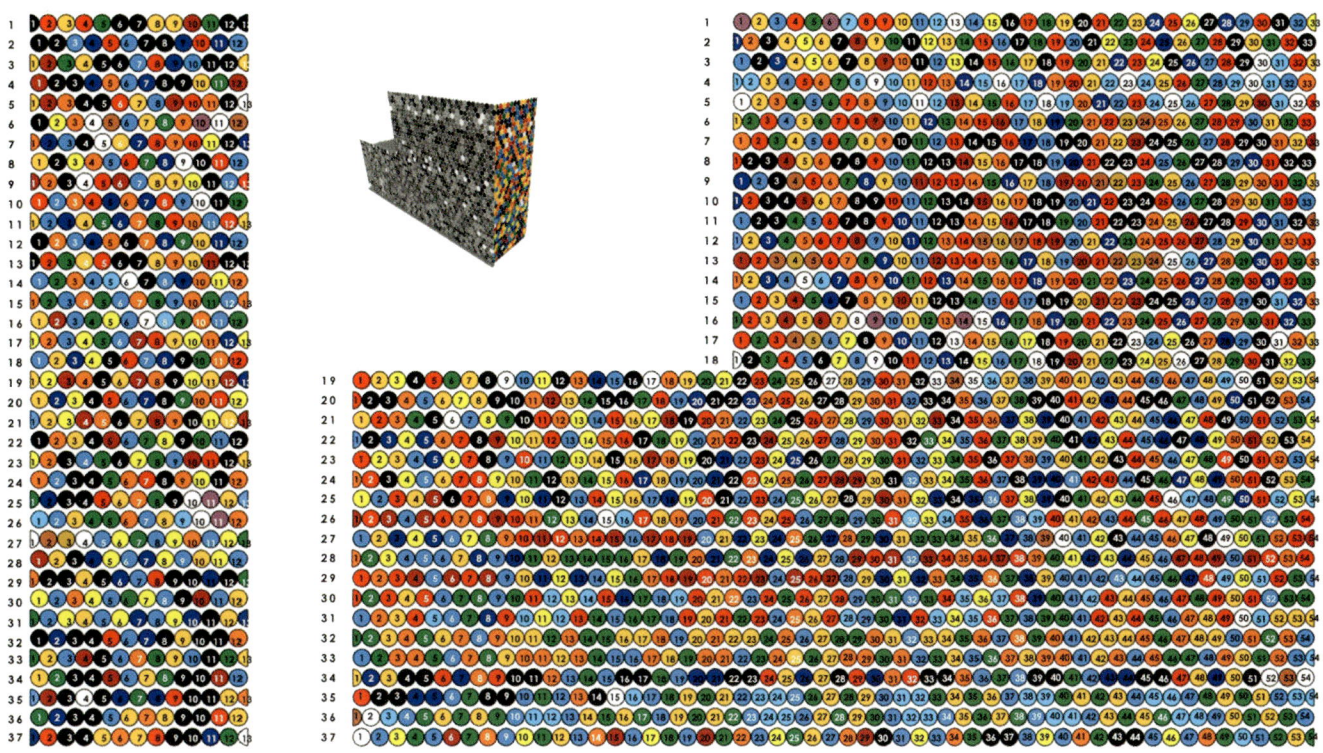

COLOR SCHEME OF FACADE

↑ Exterior view   ↓ Facade made from a drum cap

← Colorful facade panel   → Colorful facade panel

1. METALLIC STRUCTURE
2. DRAINAGE RAIN WATER TUBE
3. INCLINATED FINISHING 2%
4. 6MM IRON NET WITH 10X10MESH
5. LAYER OF WATERPROOF MEMBRANE WITH FIBRE FOR ROOF
6. SLAB
7. BEAM
8. WOODEN FRAME
9. 8MM LAMINATED GLASS
10. CONCRETE BRICKS WALL (10CM)
11. PLASTER
12. POLISHED CONCRETE
13. STEP OF STAIRCASE IN CONCRETE

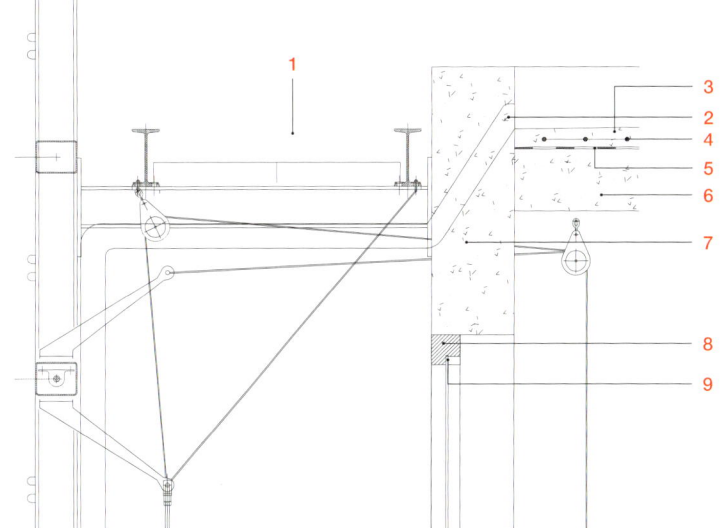
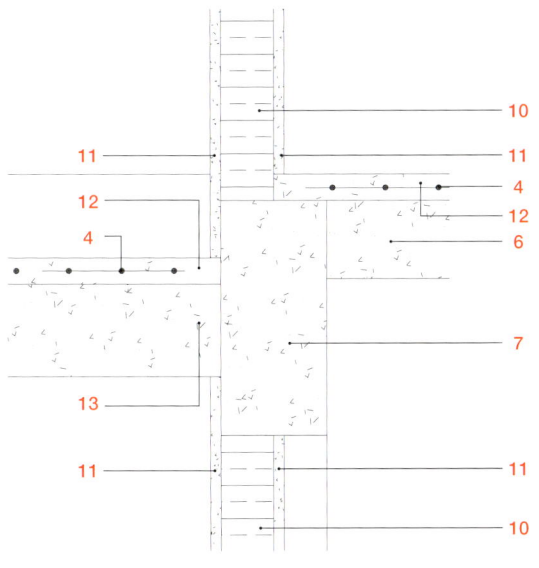
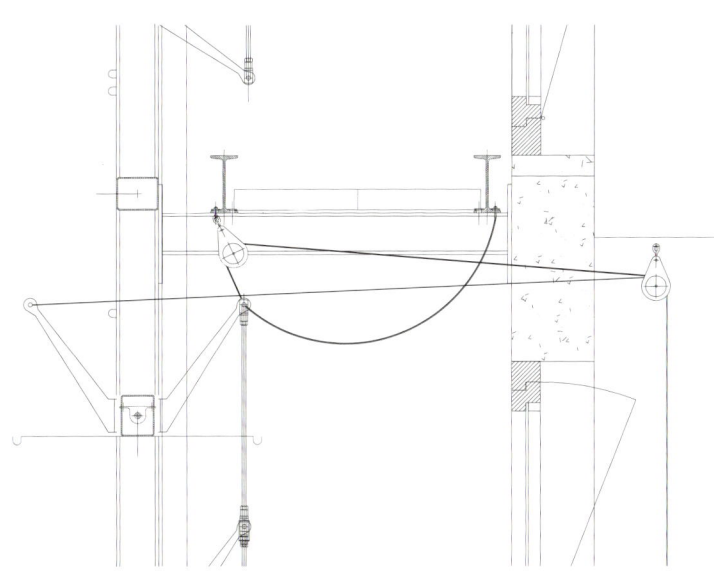

SECTION DETAIL

↑ Gallery   ↙ Gallery   ↘ Gallery

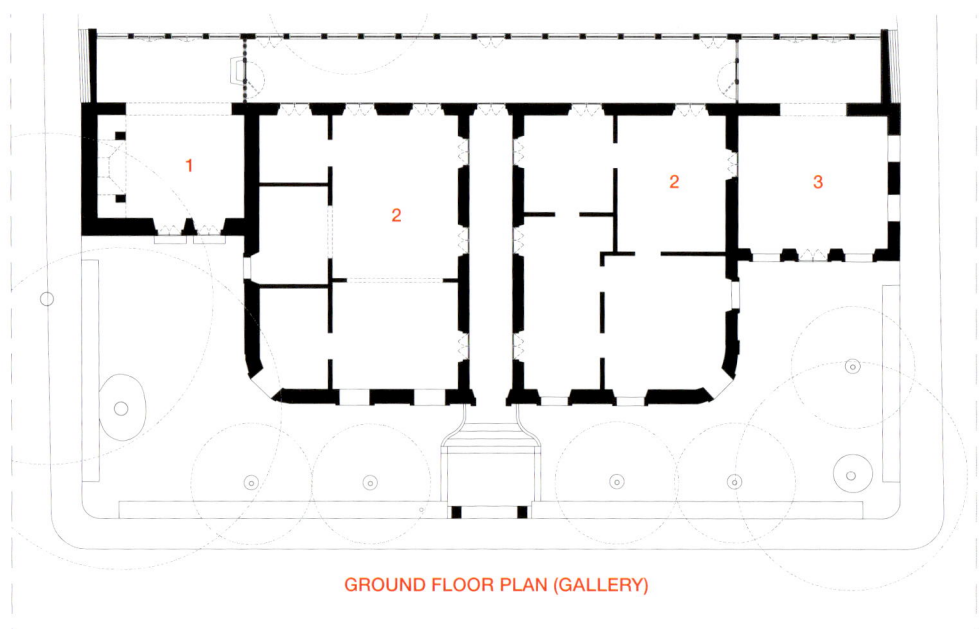

GROUND FLOOR PLAN (GALLERY)

← Library   → Library

1 ECOFFEE SHOP   3 STORE   5 COLLECTION
2 GALLERY   4 WATER WELL   6 RECEPTION

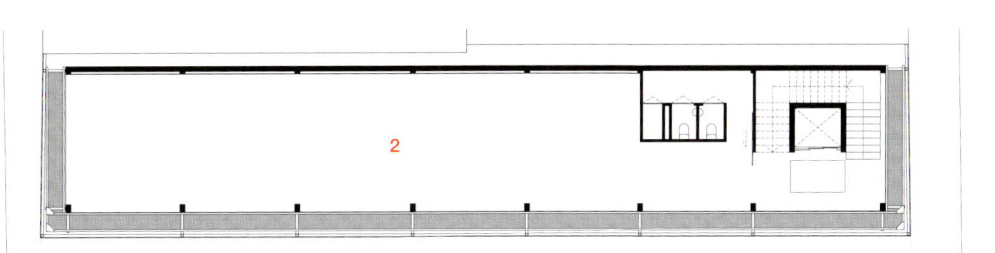

1ST FLOOR PLAN (GALLERY)

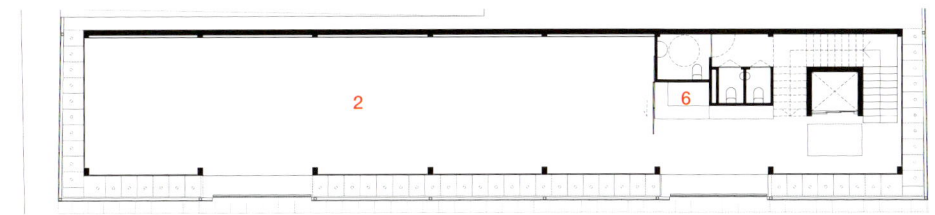

GROUND FLOOR PLAN (GALLERY)

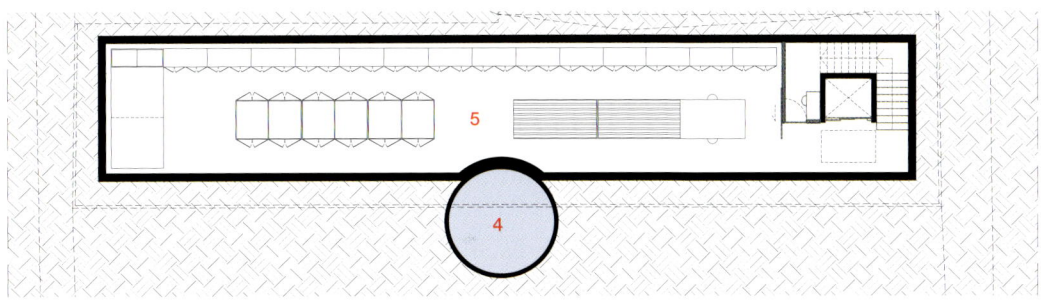

BASEMENT FLOOR PLAN (COLLECTION)

43

↑ Residence

1 BALCONY
2 LIBRARY
3 ROOM
4 CLASSROOM
5 DIRECTOR'S OFFICE
6 OFFICE
7 MEETING ROOM

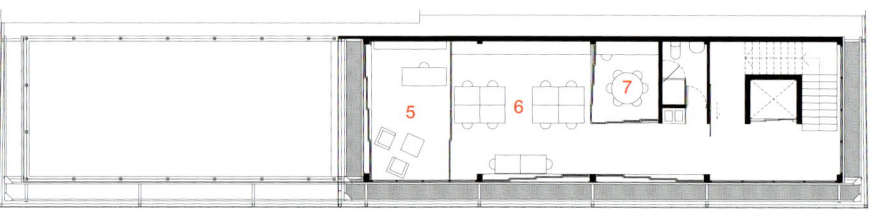

4TH FLOOR PLAN (OFFICE)

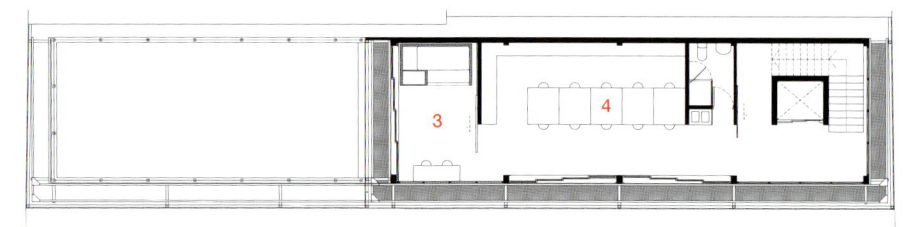

3RD FLOOR PLAN (RESIDENCE)

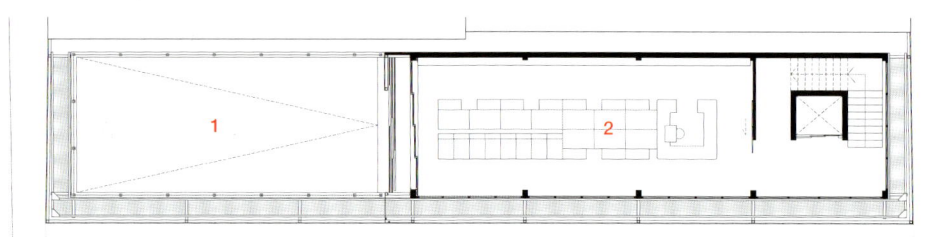

2ND FLOOR PLAN (LIBRARY)

↑ Office ↰ Balcony ↲ Exterior view → Street view

세로돌
VERTICAL STONE

ARCHITECT : YOAP ARCHITECTS LTD. / INKEUN RYOO, DORAN KIM, SANGKYONG JEONG

SEONGNAM HIGH PUBLIC HOUSING DISTRICT (hereinafter referred to as High District) is located between Inneungsan Mountain, which is connected to the foot of Cheonggyesan Mountain, and Sangjeokcheon Stream flowing from Daewang Reservoir to Tancheon Stream. The "High District" is a small residential development district with four small public housing complexes and four small detached housing blocks. Apartments have a height limit of 15 stories, and detached house lots have a floor area ratio limit of 160%, so the density is lower than other planned districts.

The block with the vertical stone house plots also consists of 26 detached house plots, so it is a housing development district, but it has the characteristics of a cozy small village. Among the four housing lots that form the entry face of the block, the lot of the vertical stone house is located in the corner and forms the facade by touching the longest street, so it was judged that it can have a great impact on the impression of the village.

I hoped that it would be a building that could have a bright effect on the landscape and image of the village when looking at the building upon entering. I imagined a building with a white image, and took an unfamiliar, awkward and impressive image reminiscent of a stone or statue rather than a house.

The opening is not visible on the facade of the building that can be seen for the first time while approaching from the corner. The building's empty windowless face on the west side rises steeply at a small width, producing an unsettling sensation of tension. I wanted the building to appear utterly abstract from some angles. The building is completely abstracted by hiding not only the openings but also the functional elements of the building that are easily exposed to the exterior, such as stormwater pipes, gas pipes, ventilation ducts, flue pipes, and outdoor units.

A slight turn at the corner reveals the hidden windows on the southern side. A vertical wall and louver on the front were planned to utilize the advantages of the land facing the long road and to compensate for shortcomings such as privacy infringement. The neighborhood living facilities on the first floor are planned almost following the shape of the site facing the road, and the residential floor is slightly twisted inward and turns around, creating a depth on the inlet side of the house. The windows on the south side, which become the main facade of the unit, are widely planned, filling up between the vertical louvers on the outside. It occupies the front of the interior space and opens with a sense of openness from the inside, but the interior space is not visible from the outside due to the sense of depth of the exterior vertical wall.

The vertical windows of the first-floor shopping district, regularly and rhythmically planned along the road, lead to vertical columns to the household access road at the east end. The household access road is separated from the parking space by vertical pillars with repeatedly standing vertical patterns, creating an image of entering the entrance through a short corridor rather than approaching through a piloti.

Location 6, Godeung-ro 1-gil, Sujeong-gu, Seongnam-si, Gyeonggi-do, Republic of Korea **Use** Multiplex Housing **Site area** 257.80m² **Building area** 153.22m² **Gross floor area** 408.54m² **Completion** 2021 **Design team** Inkeun Ryoo, Doran Kim, Sangkyong Jeong, Chaelyul Kim **Structural engineer** HANGIL structural engineering **Mechanical engineer** GM EMC **Telecom-munication equipment** GM engineering **Contractor** eden construction **Photographer** Inkeun Ryoo

← View from southwest → West facade

↑ South facade

청계산 자락과 이어진 인릉산과 대왕저수지에서 탄천으로 흐르는 상적천 사이에 '성남고등 공공주택지구'(이하 '고등지구')는 위치한다. '고등지구'는 4개의 소규모 공공주택 단지와 4개의 소규모 단독주택 블록이 있는 소규모 택지 개발 지구이다. 아파트는 15층 이하의 높이 제한이 있고 단독주택 필지는 용적률 160% 제한이 있어 다른 계획지구에 비해 밀도가 낮게 계획되어 있다.

세로돌집의 필지가 있는 블록은 26개의 단독주택 필지도 구성되어 있어 택지 개발 지구이지만 아늑한 작은 마을과 같은 분위기가 나는 특징이 있었다. 블록의 진입 면을 형성하는 4개의 주택 필지 중 세로돌 집의 필지는 코너부에 위치하며 가장 길게 가로를 접하며 입면을 형성하고 있어 마을의 인상에 많은 영향을 줄 수 있다고 판단하였다.

진입하면서 건물을 바라보았을 때 마을 풍경과 이미지에 밝은 영향을 줄 수 있는 건물이길 바랐다. 하얀 이미지의 건물을 상상했고 집이라기보다는 돌이나 조각상을 연상시키는 낯설고 어색하면서 인상적인 이미지를 잡아갔다.

코너부에서 접근하면서 처음 바라보이는 건물의 입면에는 개구부가 보이지 않는다. 건물의 서측면도 개구부 하나 없이 비워진 입면이 좁은 폭으로 높게 솟아 낯선 긴장감을 유발한다. 어떤 각도에서는 건물이 완전히 추상화되기를 바랐다. 개구부뿐만 아니라 우수관, 가스관, 환기구, 연통, 실외기 등 외관에 노출되기 쉬운 건물의 기능적인 요소들이 감춰지면서 건물은 완전히 추상화된다.

코너에서 살짝 돌아서면 숨겨져 있던 남측면의 창들이 드러난다. 도로를 길게 면하는 대지의 장점을 살리고 프라이버시 침해 등의 단점을 보완하기 위해 전면의 수직 벽체와 루버가 계획되었다. 1층의 근린생활시설은 도로에 면한 대지의 형상을 거의 따라서 계획되고 주거 층은 살짝 안쪽으로 비틀어져 돌아서며 주택 주입면에 깊이가 만들어진다. 세대의 주 입구이 되는 남측면의 창들은 외부의 수직 루버 사이를 가득 채우며 넓게 계획되어 있다. 내부 공간의 전면을 차지하며 내부에서는 개방감 있게 열리는데 외부 수직 벽체의 깊이감으로 인해 외부에서는 내부 공간이 보이지 않는다. 도로를 따라 규칙적이면서 리드미컬하게 계획된 1층 상가의 수직 창들은 동쪽 끝에 세대 진입로로의 수직 기둥으로 이어진다. 세대 진입로는 반복적으로 서있는 수직 패턴 외장의 수직 기둥들로 주차공간과 분리되면서 필로티를 통해 접근하는 느낌이 아니라 짧은 회랑을 통해 입구로 진입하는 이미지를 연출하였다.

↑ Street view

DESIGN CONCEPT

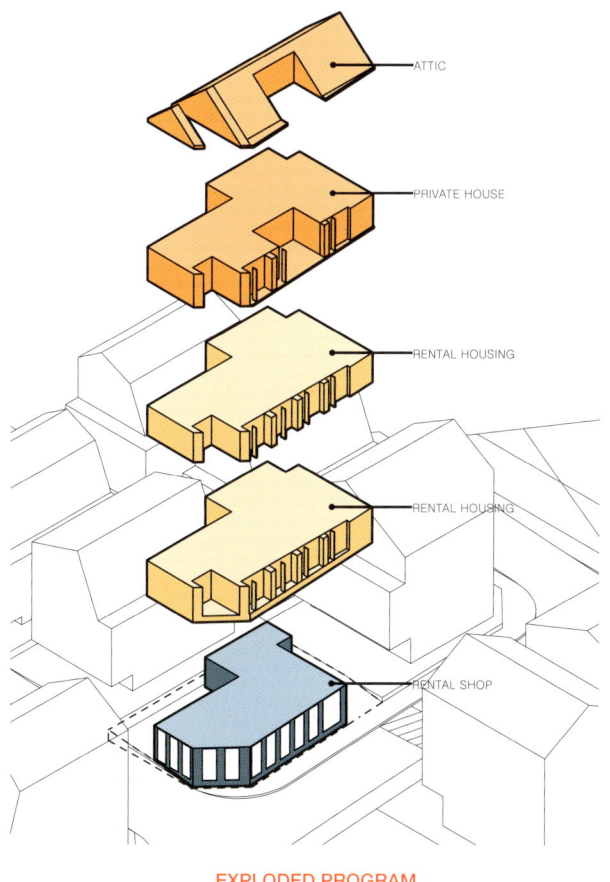

EXPLODED PROGRAM

↖ Exterior view at sunset

↙ Piloti

SOUTH ELEVATION

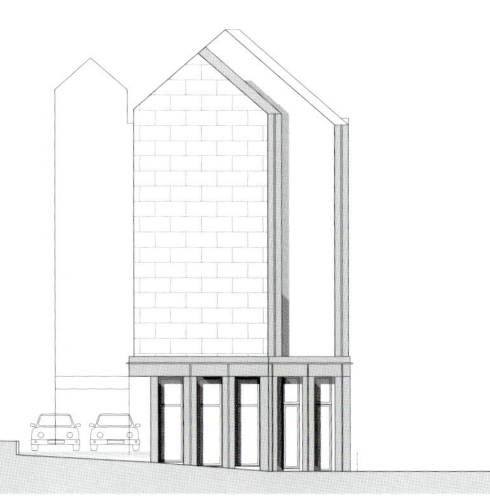

WEST ELEVATION

↑ Neighborhood facility

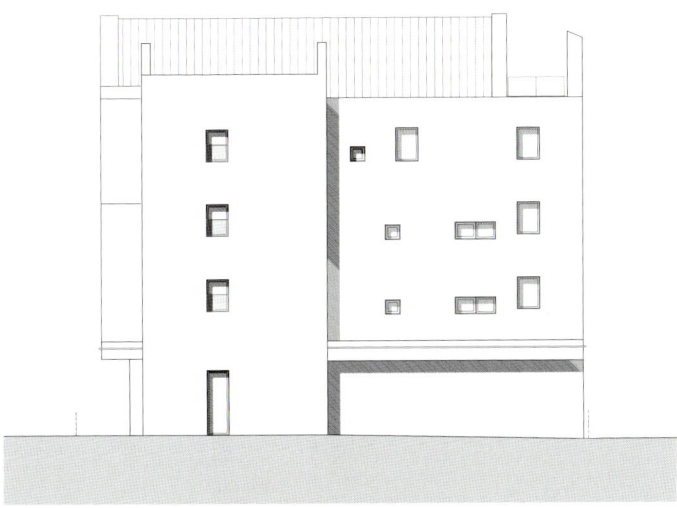

NORTH ELEVATION

EAST ELEVATION

↖ Gable roof   ↗ Partial view of exterior   ↓ Piloti

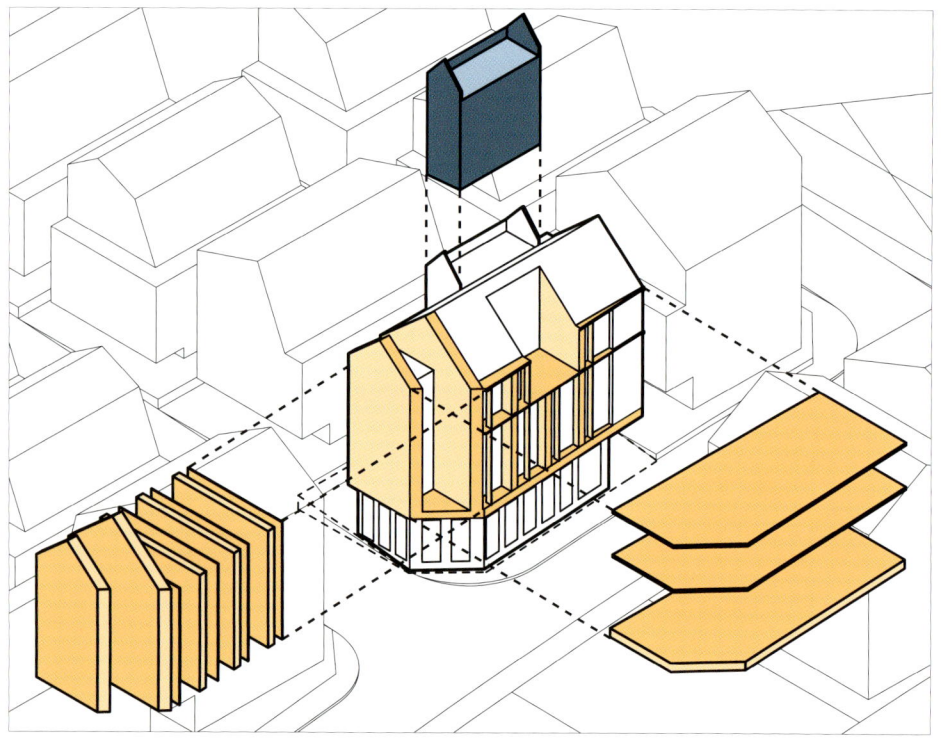

MASS SLICE

↑ Interior view   ↲ Hallway   ↳ Foyer

↑ Interior view   ↙ Tembarboard wall   ↱ Entrance   ↳ Wood partition

1	NEIGHBORHOOD FACILITY	4	BATHROOM	7	KITCHEN	10	PARKING LOT
2	BEDROOM	5	VERANDA	8	DRESS ROOM	11	LIVING ROOM
3	HALLWAY	6	ATTIC	9	RESTROOM	12	LIFT

↱ Veranda
→ Staircase
↳ Staircase

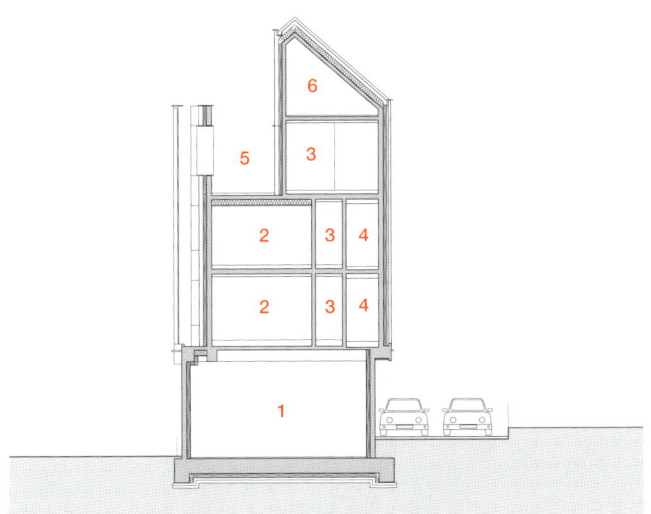

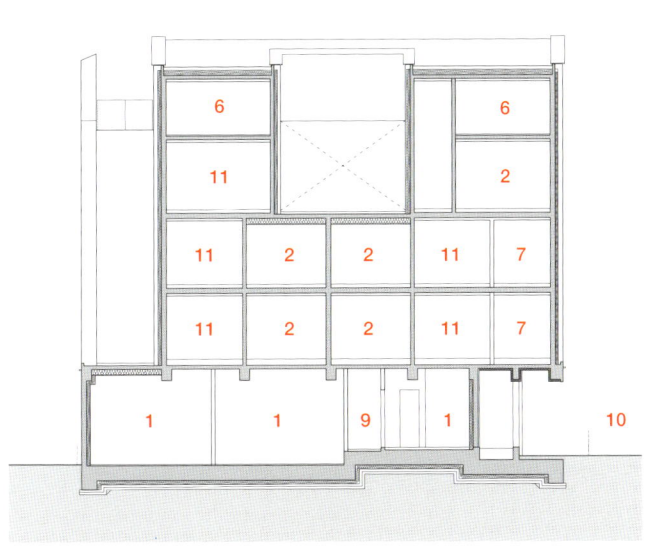

SECTION

← Interior and street view   → Bedroom

1 ENEIGHBORHOOD FACILITY
2 LOBBY
3 RESTROOM
4 PARKING LOT
5 LANDSCAPE SPACE
6 BEDROOM
7 VERANDA
8 KITCHEN & DINING ROOM
9 LIVING ROOM
10 BATHROOM
11 FOYER
12 ELEVATOR HALL
13 DRESS ROOM
14 ATTIC

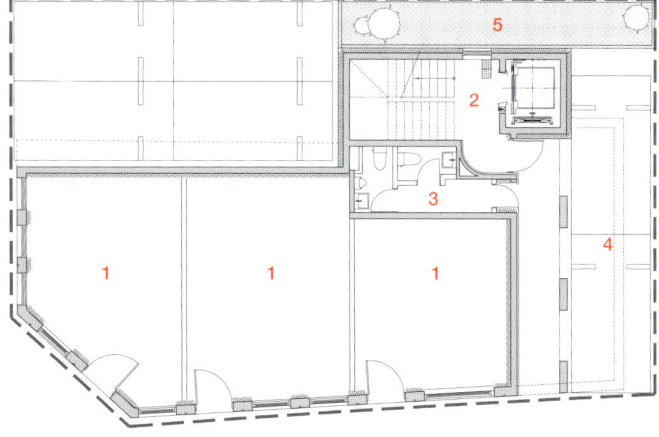

1ST FLOOR PLAN

2ND FLOOR PLAN

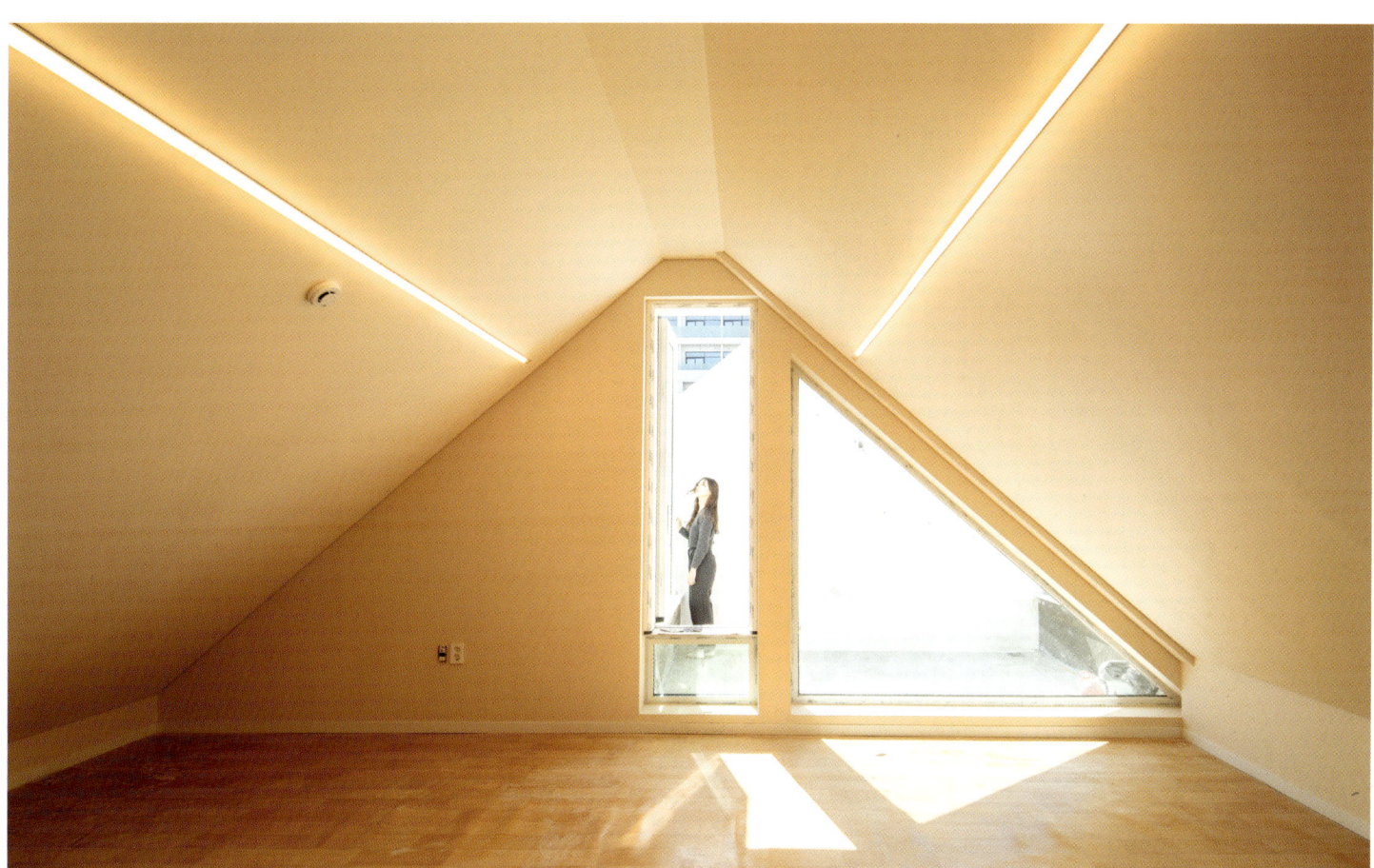

↑ Attic

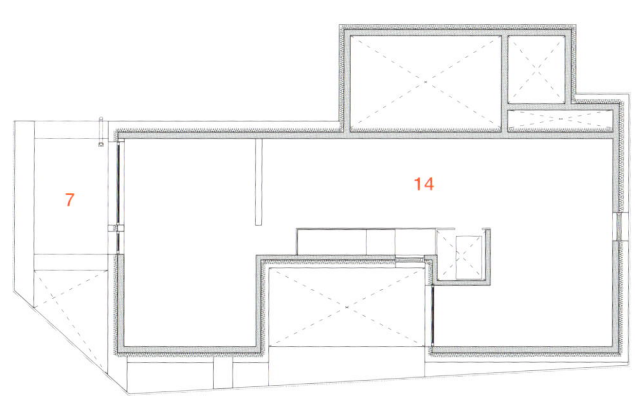

5TH FLOOR PLAN

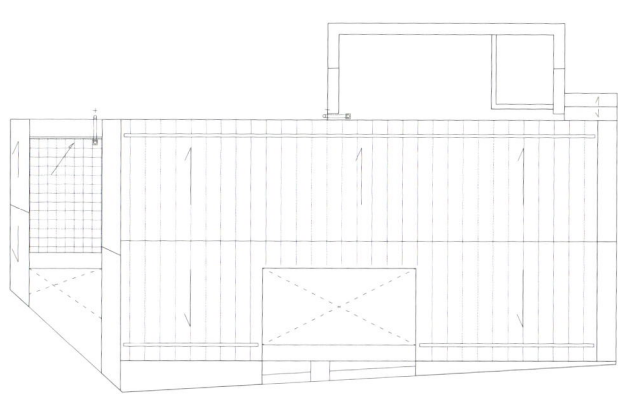

ROOF PLAN

3RD FLOOR PLAN

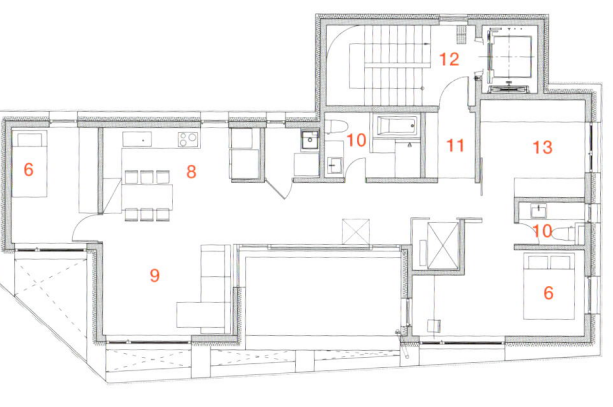

4TH FLOOR PLAN

그레미오 학교
ESCOLA GRÉMIO

ARCHITECT : FALANSTÉRIO ATELIER DE ARQUITECTURA / RICARDO DIAS, BRUNO TINOCO

FOR THE EXPANSION OF THE SCHOOL, THE GOAL WAS TO BUILD A LIGHT STRUCTURE that would not overload the existing one. Green roofs, not accessible and requiring low maintenance, were chosen to minimize the visual impact and improve the thermic conditions of the rooms through passive systems.

The proposal for the new spaces aimed for serenity and harmony, focusing on rounded shapes, reducing angles and edges, and using a palette of pastel colors. Through this approach, the space inspires calm, allowing for socialization and inclusion and promoting collaborative activities and spirits.

The chosen materials guarantee ideal thermal and acoustic comfort and ensure aesthetic and constructive quality. A light metal structure was chosen for the walls, using the capoto system and OSB. The interior insulation incorporates an air box, rock wool and double plasterboard.

The same applies to the construction of the roof, which consists of a landscaped layer of pebble, waterproofing screen, OSB board, metal structure with an air box, rock wool insulation and plasterboard on the inside.

By slightly setting back the remaining rooms the impact of the new construction on the adjacent land was greatly reduced. Openings to the south were also created with ventilation slots to improve air quality and create cross ventilation, increasing classrooms salubrity.

In this context, the school expansion added a new floor with 6 rooms, sanitary facilities, and a small storage in the south part of the school, never exceeding the maximum height of its north part. All rooms are interconnected by a canopy that ensures students are not exposed to the weather, allowing the outside space to be protected.

Location Campo de Ourique - Lisbon, Portugal **Use** Education **Built area** 526m² **Gross floor area** 526m² **Completion** 2022 **Project manager** Ricardo Dias **Design team** Bruno Tinoco, Vera Coimbra, Paulo Lima **Contractor** Grémio – Instrução Liberal de Campo de Ourique + CCE. Lda **Photographer** João Guimarães

WEST ELEVATION

NORTH ELEVATION

← Exterior view

↑ The building extension, new floor with 6 rooms

NORTH ELEVATION DETAIL

1. EXTERIOR WALLS IN METALLIC STRUCTURE WITH CAPOTTO SYSTEM. FORESEE A REINFORCEMENT OF THE MESH OF THE CAPOTO SYSTEM WITH A NET UP TO THE HEIGHT OF THE ENTRANCE SPACES, TO GUARANTEE ROBUSTNESS. COLORS OF THE ROOMS IN PASTEL TONES:
   1.1 - ROOM 1 - BLUE NCS: S 2020-B   1.2 - ROOM 2 - GRAY NCS: S 2005-R80B   1.3 - ROOM 3 - YELLOW NCS: S 1010-Y20R   1.4 - ROOM 4 - GREEN NCS: S2020-B70G   1.5 - ROOM 5 - GRAY NCS:PE.2.01   1.6 - ROOM 6 - ORANGE NCS: S1030-Y0R   1.7 - SPACES BETWEEN ROOMS - NCS: 26502-B
2. WALL IN METALLIC STRUCTURE WITH THERMAL COATING IN CAPOTTO SYSTEM IN WHITE RAL 9010
3. PAINTING ROOM IDENTIFICATION NUMBERS WITH REFLECTIVE ACRYLIC PAINT IN THE COLOR
4. GUARD IN FLAT IRON BAR AND CIRCULAR RODS WITH EPOXY PAINT FINISH IN WHITE RAL 9010
5. EXTERIOR FLOOR OF THE ACCESS CORRIDOR IN NON-SLIP CERAMIC MOSAIC OR MORE IDENTICAL TO THE EXISTING MOSAICS ON THE REMAINING FLOORS
6. ALUMINUM FRAMES WITH THERMAL BREAK AND DOUBLE GLAZING WITH AIR BOX, INCLUDING OUTER FRAME IN BENT SHEET METAL. FIXED AND MOVABLE FRAME WITH LACQUERED FINISH IN GRAY COLOR RAL 7046 EXTERIOR FRAME IN BENT METAL SHEET PAINTED IN RAL COLOR 7046
7. PROTECTIVE VISOR IN METALLIC STRUCTURE COATED WITH BENDED METALLIC SHEET ON THE SIDES FINISHED IN RAL 9010 WHITE PAINT AND LOWER FINISH IN AQUAPANEL IN WHITE RAL 9010. FINISHING AREAS BETWEEN THE BUILDING AND THE CANOPY WITH ACRYLIC HIGH-RESISTANCE COLORLESS WITH SLOPE FOR RAINWATER DRAINAGE
8. PAVEMENT IN THE TECHNICAL AREA COATED WITH WATERPROOFING AND INSULATING FABRIC WITH APPLICATION OF GEO-TEXTILE BLANKET, COVERED BY GRAVEL UP TO 7cm
9. COVER TRIM BETWEEN ROOMS IN PRE-AGED ZINC PLATE
10. LADDERS IN METALLIC STRUCTURE; PAINTED OLIVE LEAF TYPE SHEET METAL STEPS IN WHITE RAL 9010 EXTERIOR FINISH IN SHEET METAL PAINTED IN RAL 9010
11. EXISTING METALLIC GUARD REPAIRED AND PAINTED IN THE SAME GRAY COLOR AS THE EXISTING ONE

→ Rounded shapes & palette of pastel colors

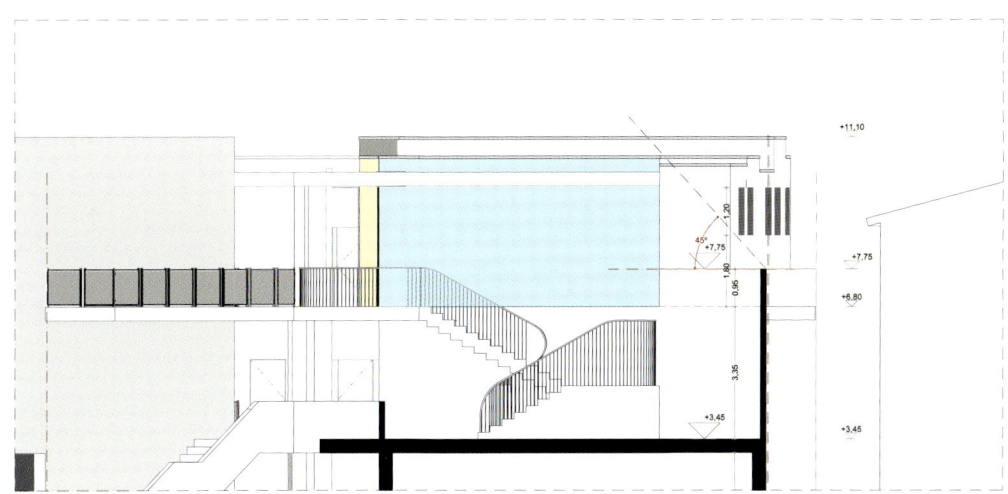

WEST ELEVATION DETAIL

학교를 확장하는 작업의 목표는 기존 건물에 과부하가 걸리지 않도록 가벼운 구조를 만드는 것이다. 접근성이 없고 유지 관리 비용이 적은 녹색 지붕을 선택해 시각적 영향을 최소화하고 폐쇄적 구조의 특성을 활용해 실내 열 조건을 개선하고자 하였다.
각도와 가장자리를 줄여 최대한 곡선형 형태를 살리고 파스텔 톤 색상을 더함으로써 편안하고 조화로운 분위기의 공간으로 재구성하고자 했다. 이로써 사회화와 포용력 있는 태도를 고취하고 협력 활동과 정신을 고양할 수 있는 차분한 분위기를 조성했다.
여기에 이상적인 온도 조절 및 편안한 방음을 보장하는 실용적이면서도 미적인 재료를 선택했다. 벽은 카포토 시스템과 OSB를 적용해 가벼운 금속으로 구성했다. 내부 설치물은 에어박스와 암면 및 이중 석고 보드로 마감했다.

이는 천장에도 동일하게 적용되어 조경을 위한 자갈층과 방수 스크린, OSB 보드, 에어 박스와 금속 구조물, 암면 단열재, 내부의 석고보드로 마감했다.
남은 공간을 약간 뒤에 배치해 신축 공사로 인한 인접 지역의 피해를 크게 줄였다. 남쪽으로 개방된 곳에 환기 시설을 추가해 공기 질을 개선하고 교차 환기를 통해 교실의 환경을 한층 쾌적하게 개선했다.
이러한 기조를 유지하며 학교 건물 남쪽에 6개의 공간과 위생 시설, 작은 창고를 북쪽의 최대 높이를 넘지 않는 선에서 추가로 구성해 확장했다. 모든 공간은 학생들이 가혹한 날씨에 노출되지 않도록 캐노피로 연결되어 외부 공간을 보호할 수 있다.

↑ Playground & west side
↵ Rounded shapes canopy

↑ Rounded shapes canopy & balcony   ↙ Rounded shapes canopy & pastel colors   ↘ Rounded shapes canopy & pastel colors

↑ Rounded shapes canopy & balcony   ↙ Stair   ↳ Classroom

1 GRAVEL
2 WATERPROOFING SCREEN
3 2cm OSB BOARD
4 METAL STRUCTURE IN 25X5cm PROFILE
5 METALLIC PROFILE FOR FIXING THE FLASHING AND DRIP TRAY SUPPORT
6 METALLIC FLASHING
7 MECHANICAL FASTENING
8 1.8cm OSB BOARD
9 MECHANICAL FASTENING
10 AIR BOX
11 12cm ROCK WOOL INSULATION
12 60mm INSULATION BOARD
13 COATING REINFORCED WITH MESH
14 FINISHING LAYER
15 DOUBLE 1.25cm PLASTERBOARD
16 7cm ROCK WOOL INSULATION
17 STRUCTURE IN METALLIC PROFILE OF 20X5cm
18 DRIP PAN
19 HIGH STRENGTH COLORLESS ACRYLIC SHEET
20 SHEET METAL FRAME IN RAL 7046 COLOR
21 RAILING IN SHEET METAL IN WHITE RAL 9010
22 HOOD FINISH IN SHEET METAL IN WHITE RAL 9010
23 EXISTING WALL REPAIRED AND PAINTED IN WHITE RAL 9010
24 WATERPROOFING
25 TOPSOIL
26 EXISTING REINFORCED CONCRETE SLAB
27 RESILIENT BLANKET
28 LEVELING SCREED
29 ROLL VINYL FLOORING
30 HALF-ROUND VINYL BASEBOARD
31 SILL IN SHEET METAL IN RAL 7046 COLOR

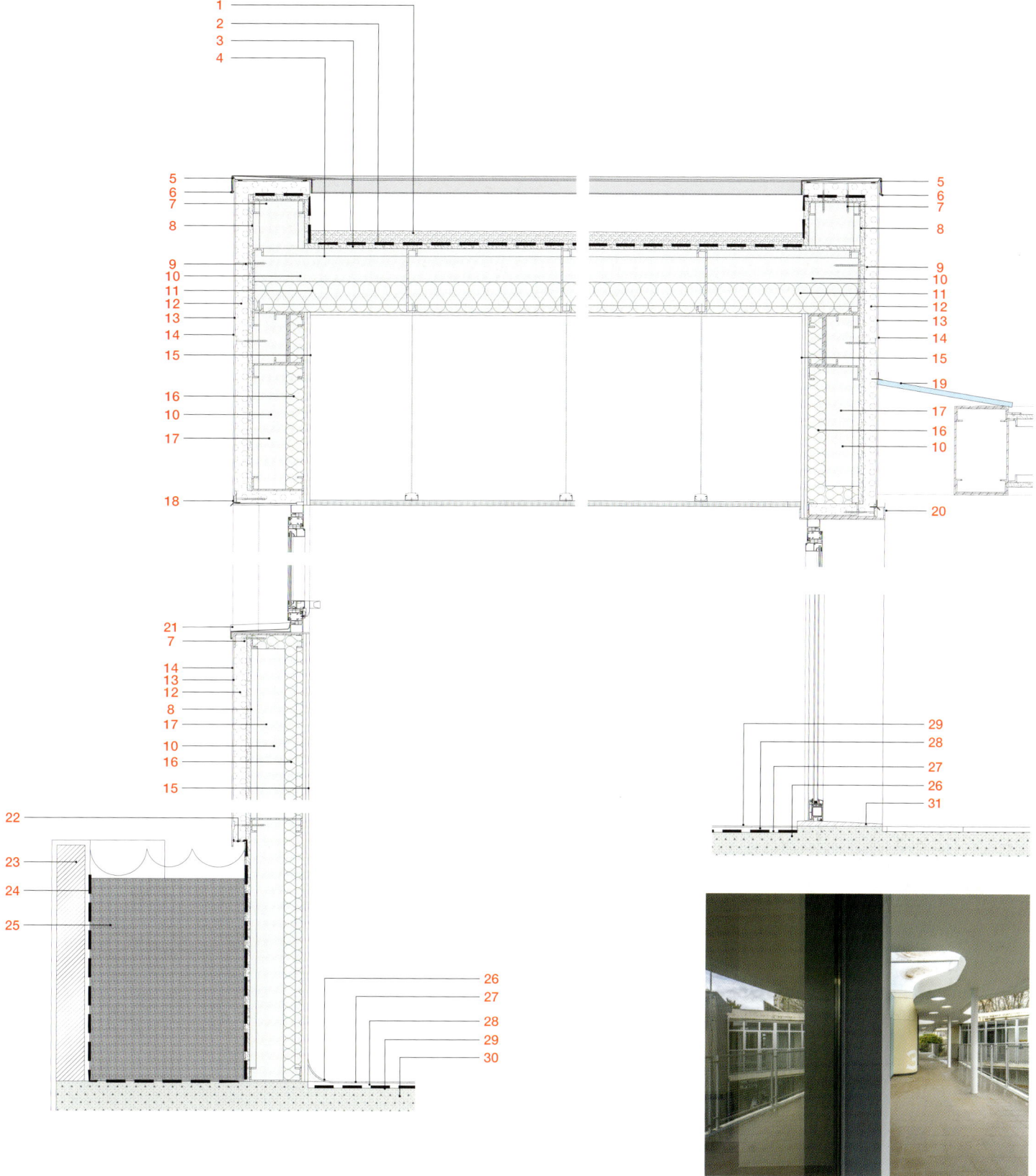

DETAIL

↑ Rounded shapes canopy & classroom

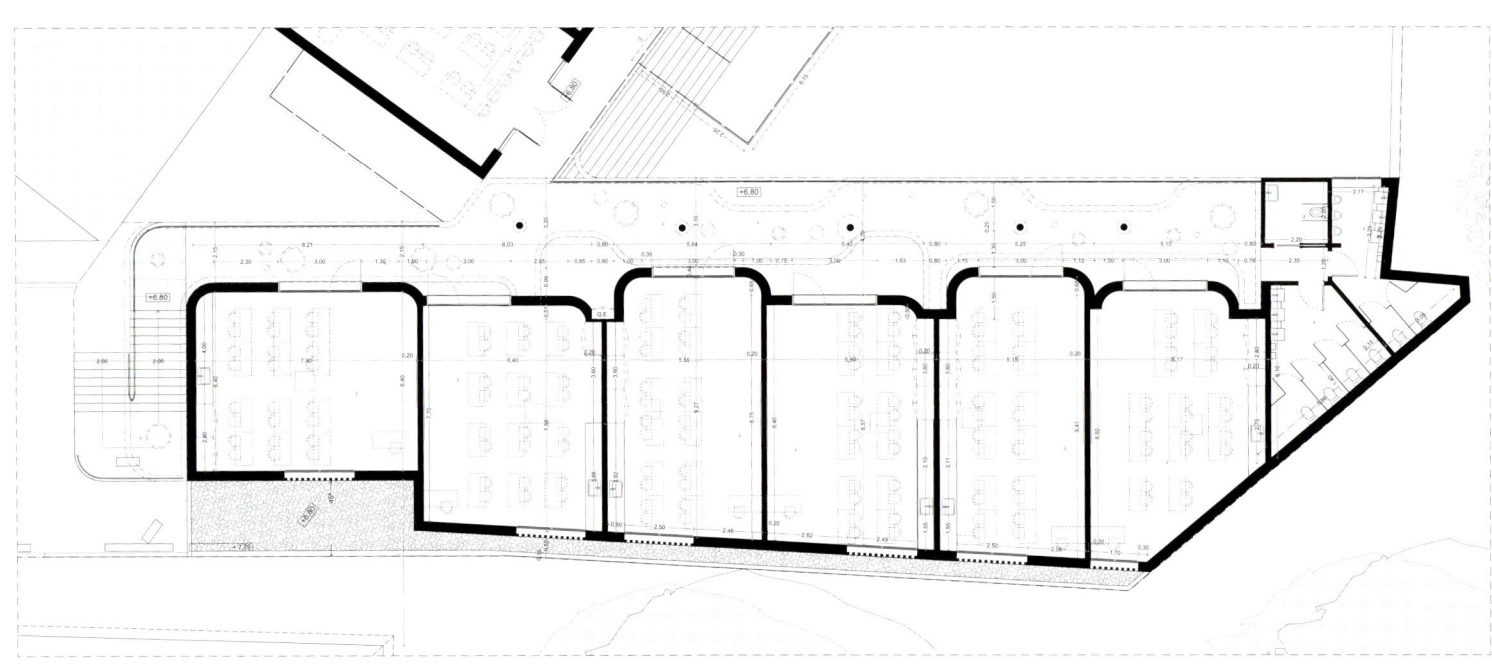

FLOOR PLAN

← Room → Room

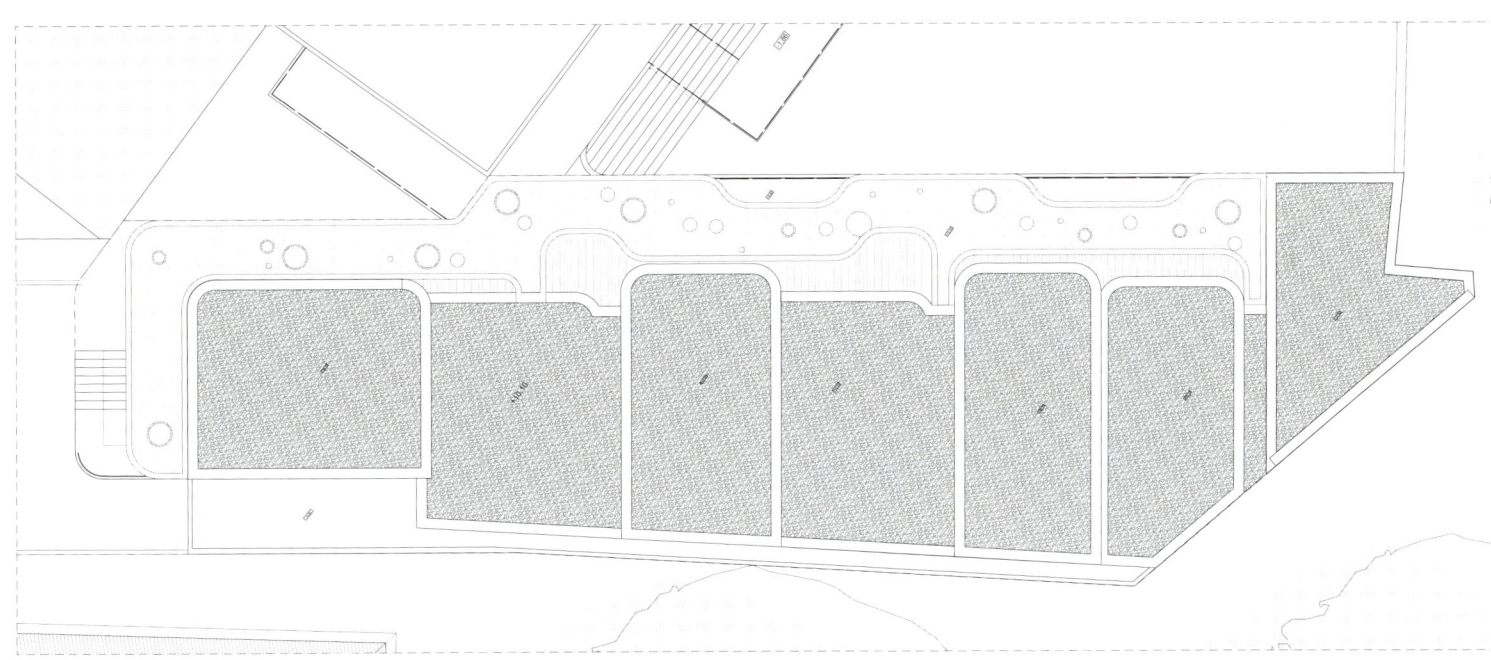

ROOF PLAN

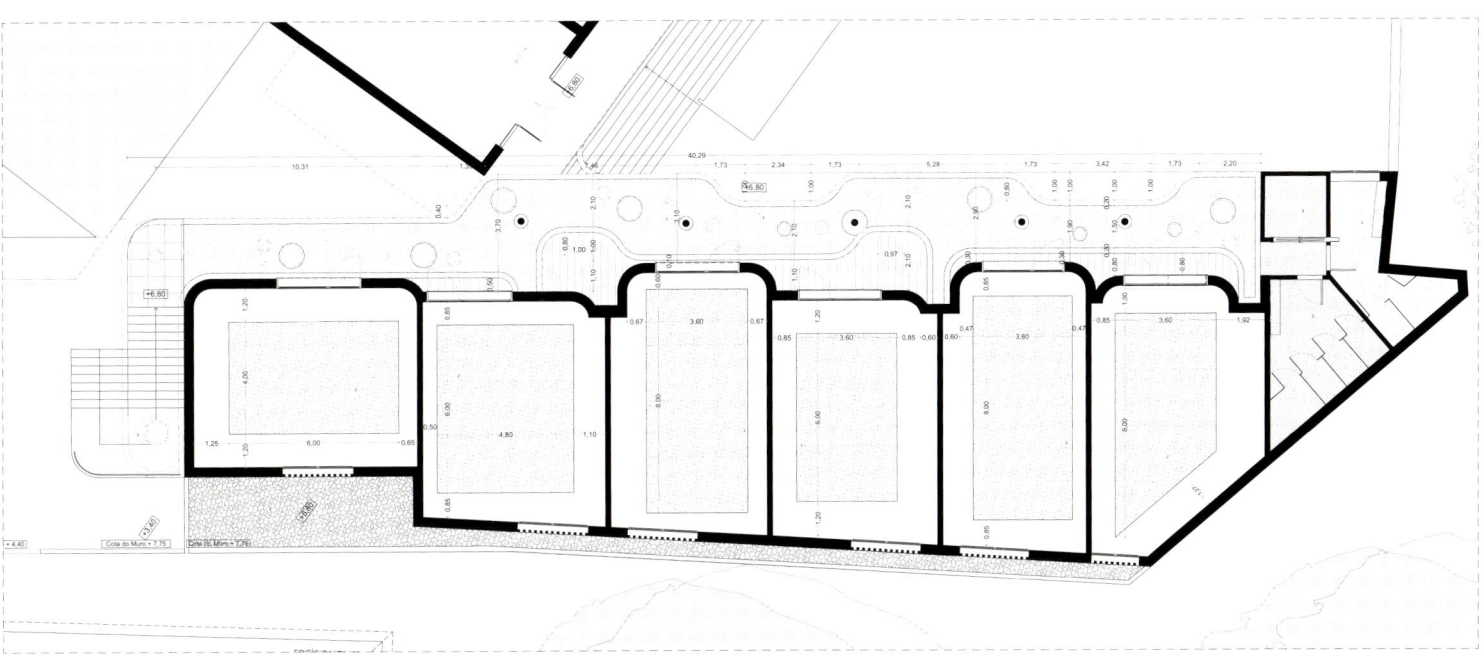

CEILING PLAN

라 리나센테 – 피우메 광장

LA RINASCENTE – PIAZZA FIUME

ARCHITECT : 2050+ / IPPOLITO PESTELLINI LAPARELLI

2050+ HAS COMPLETED THE ARCHITECTURAL TRANSFORMATION of the exteriors and the refurbishment of the sixth floor of La Rinascente in Piazza Fiume, Rome, one of the most iconic buildings designed by Franco Albini and Franca Helg between 1957 and 1961.

The timeless modernity of the two masters of Italian architecture, which does not rest on style or image, but on materials and function, is translated into a series of targeted interventions that took on the legacy of the original project.

The interventions are grounded on tactical rather than formal considerations. They include the strategic replacement of specific building components with the aim to revive both technical and spatial solutions that were once part of the original project, while also adapting La Rinascente in Piazza Fiume to the requirements of a contemporary department store.

The newly built panoramic elevator, which replicates the form of the stairwell as designed in the 1957 version of the project, is treated as a technical insert inside the structural partition of the façade along the courtyard: it flanks the existing egress stairs, improving internal flows while providing direct access to the restaurant on the sixth floor, directly from the courtyard, also outside opening hours. All the shop windows alongside Piazza Fiume and via Salaria have been replaced, restoring original geometries and proportions. The transparency of the courtyard glass surfaces has also been restored, revealing once again the escalators and rehabilitating their scenographic function. The portion of the roof facing the Aurelian Walls is transformed through the introduction of a glass surface that offers visitors an unprecedented view of the city from the restaurant on the sixth floor.

The opening of the sixth floor marks the end of the restoration of the façade and the completion of the new architectural project that strives to re-establish the building as an open space for citizens and visitors alike. Seeking a balance between restoration and transformation, the project by 2050+ aims to build continuity with the work of Albini and Helg, while developing a dynamic relationship with the past.

Location Piazza Fiume, Rome, Italy **Use** Retail **Completion** 2022 **Design team** Giacomo Ardesio, Mattia Inselvini, Francesca Lantieri, Camilla Morandi, Ippolito Pestellini Laparelli, Massimo Tenan, Chiara Tomassi **Contractor** CMB **Photographer** DSL Studio – Agnese Bedini, Alessandro Saletta

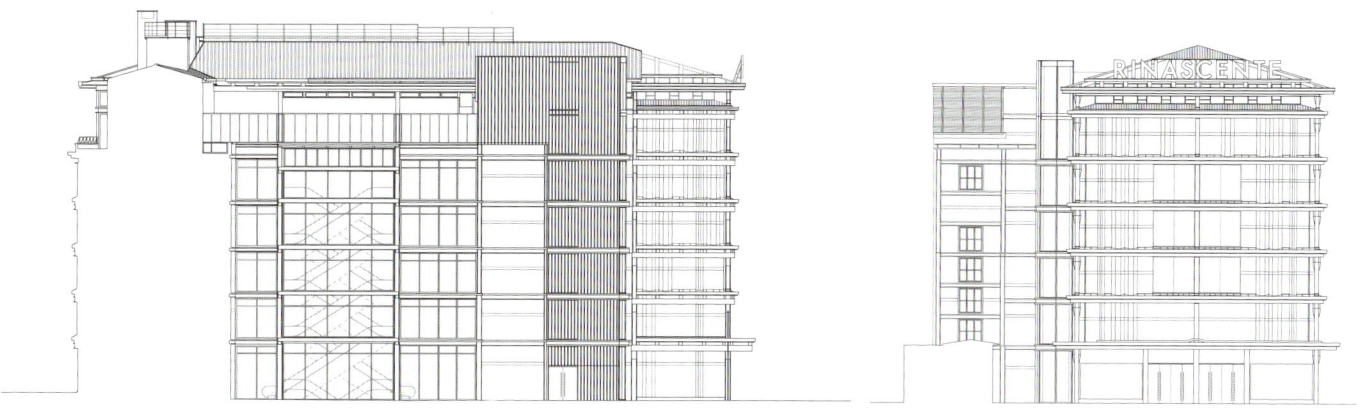

ELEVATION

← Exterior view

← Exterior view
↱ Exteriors & remodeling view
↳ Exteriors & remodeling view

2050+는 1957~1961년에 프랑코 알비니와 프랑카 헬그가 설계한 가장 상징적인 로마의 건물 중 하나인 피아자 피우메의 라 리나센테 6층 외부 개조 및 보수 공사를 진행한 프로젝트이다. 스타일이나 이미지가 아니라 재료와 기능을 중시하는 이탈리아 건축의 두 거장의 시대를 초월한 모던한 감각이 깃든 기존 프로젝트의 유산을 전수하면서도 일부 재해석을 시도하였다. 이때 형태보다는 전략적인 부분에 중점을 두었다. 그래서 원래 건물의 기술 및 공간적 요소를 되살리면서도 피아자 피우메의 라 리나센테를 현대적인 백화점으로 변화시키려는 요구사항을 충족시키고자 전략적으로 건물의 일부 요소를 교체했다.

안뜰을 따라 구조적으로 분할된 파사드 내부에 1957년 설계된 계단 통의 형태를 복원한 파노라마 엘리베이터를 기존 출구 계단 측면에 새로 구축해 동선 흐름을 개선하면서도 영업시간이 지난 후에는 안뜰에서 6층 레스토랑으로 바로 이동할 수 있게 하였다. 피아자 피우메를 따라 살라리아로 이어지는 상점 창문도 전부 교체해 본래의 기하학적인 비율로 복원했다. 안뜰 창 표면의 투명도 복원과 에스컬레이터를 들어내 씨노그래피적 공간으로 복원시켰다. 아우렐리아누스 방벽과 마주 보는 지붕 부분에는 유리창을 설치해 방문객들이 6층 레스토랑에서 멋진 전망을 즐길 수 있게 하였다.

6층을 개방하는 것은 시민과 방문객 모두를 위해 열린 공간으로 재정립하기 위해 파사드 복원 및 새로운 건축 프로젝트를 성공적으로 완료한 것을 의미한다. 복원과 변형의 균형에 중점을 둔 2050+ 프로젝트의 목표는 알비니와 헬그의 유산을 전수하면서도 과거를 더욱 역동적으로 발전시키는 것에 있다.

↑ The builing night view & escalators

DIAGRAM

↑ The newly built panoramic elevator   ↙ The newly built panoramic elevator   ↓ The newly built panoramic elevator   → The newly built panoramic elevator

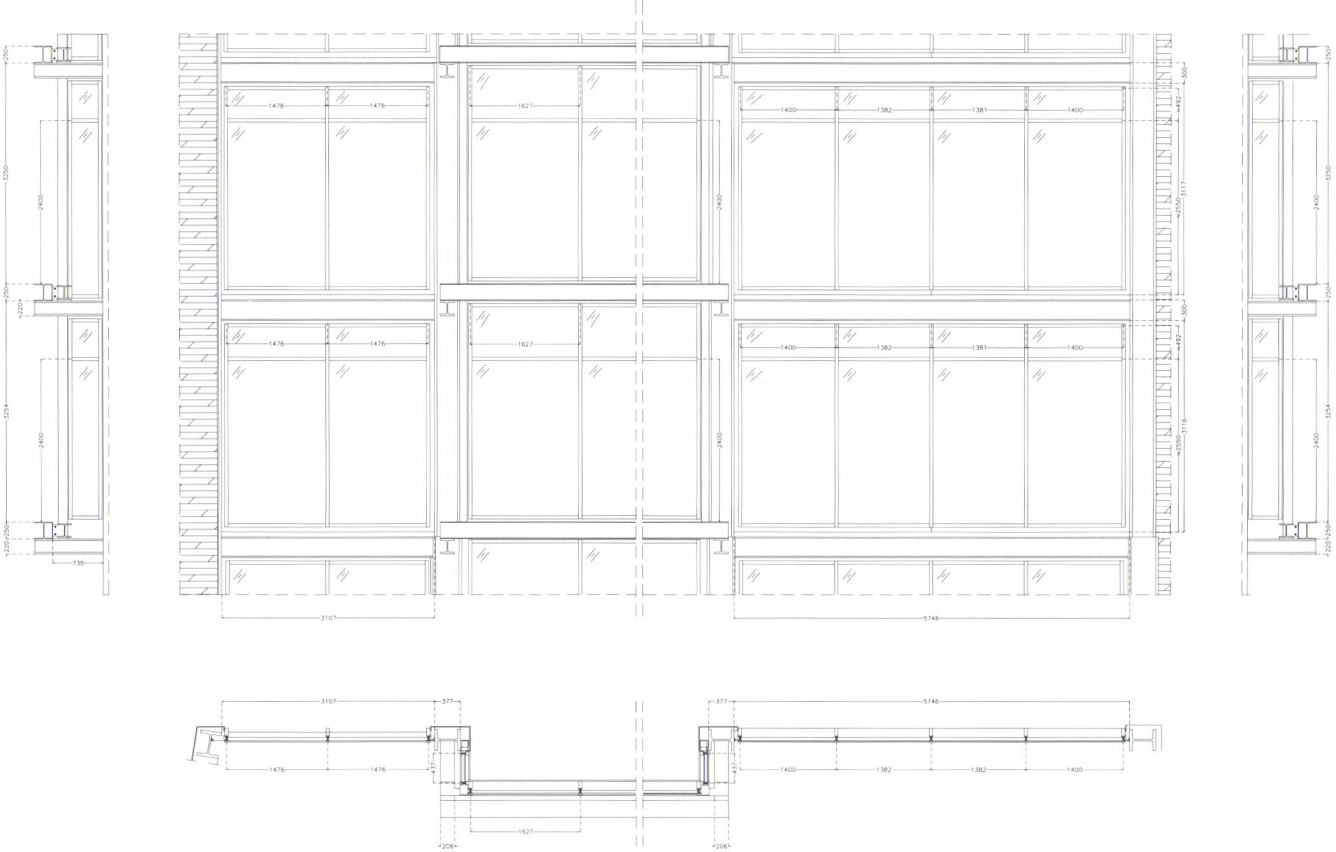

DETAILS OF THE GLAZED FACADE

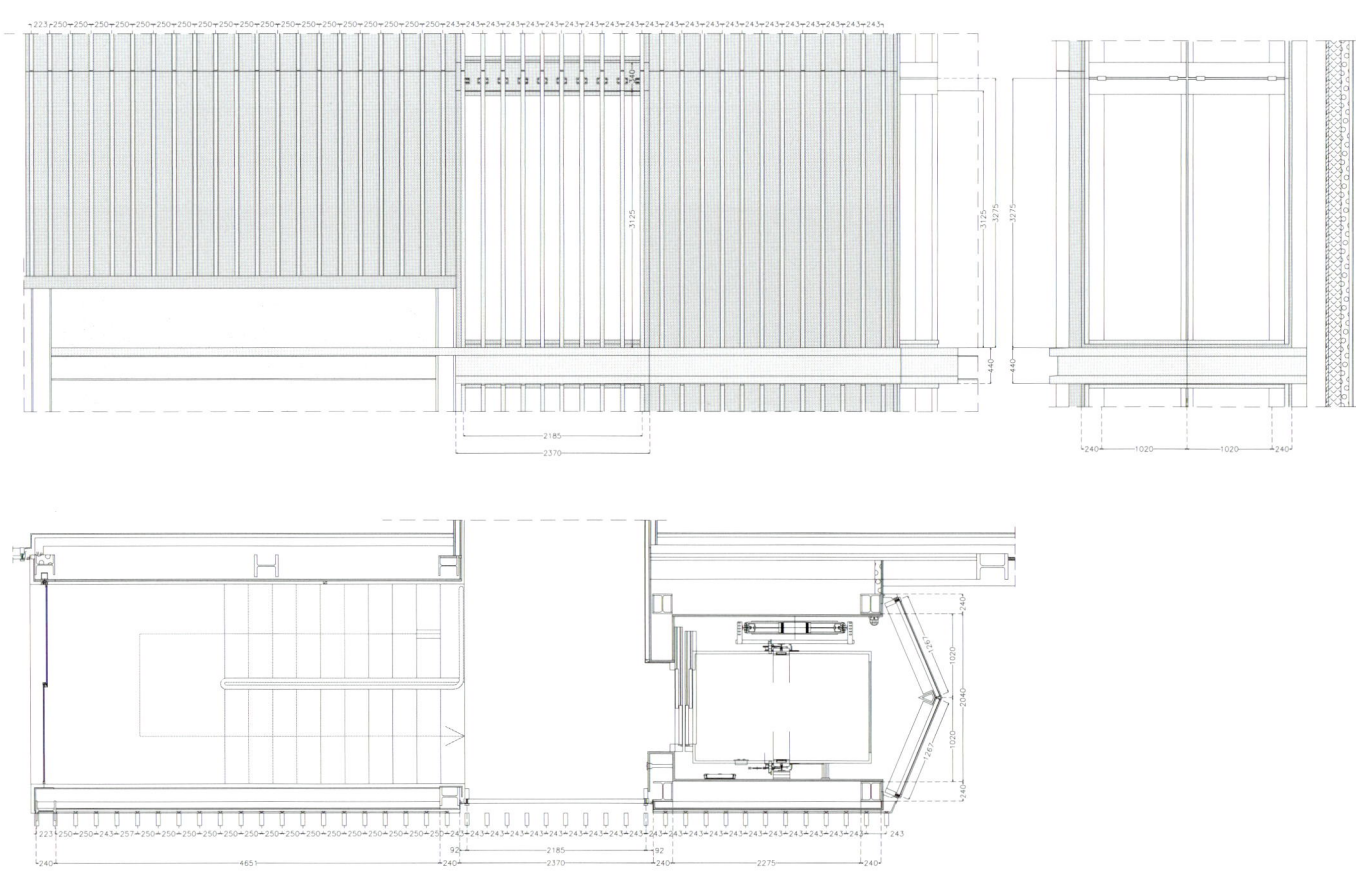

ELEVATION, SECTION & TYPE PLAN

↑ 6th floor restaurant

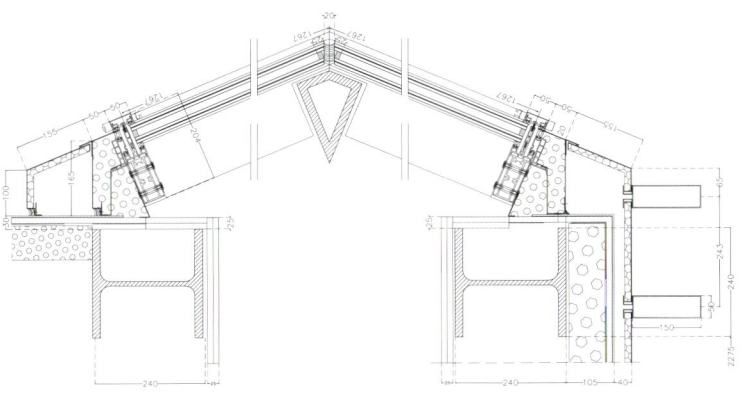

HORIZONTAL SECTION DETAIL – CUSP & LIFT NODE

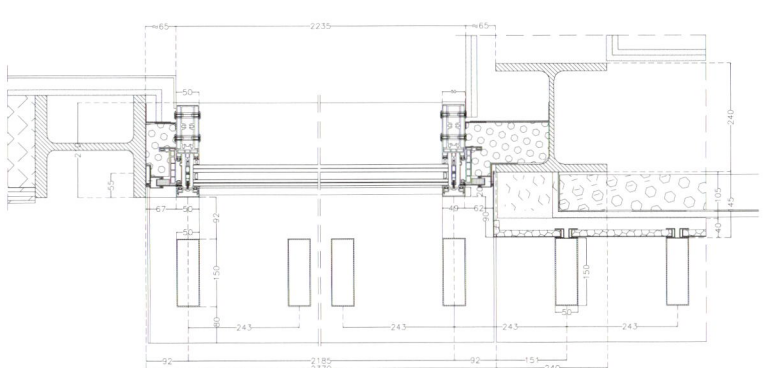

HORIZONTAL SECTION DETAIL – ELEVATION TOWARDS THE INTERNAL COURTYARD

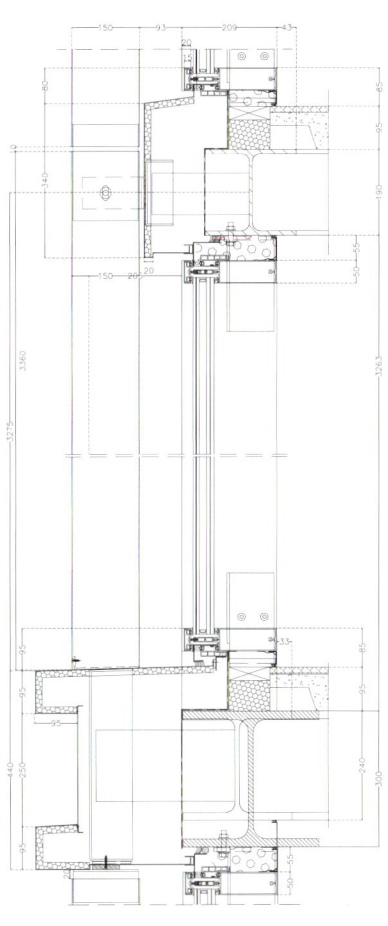

VERTICAL SECTION DETAIL – ELEVATION TOWARDS THE INTERNAL COURTYARD

↑ 6th floor shop   ← Stair   ↓ The newly built panoramic elevator   → Stair detail

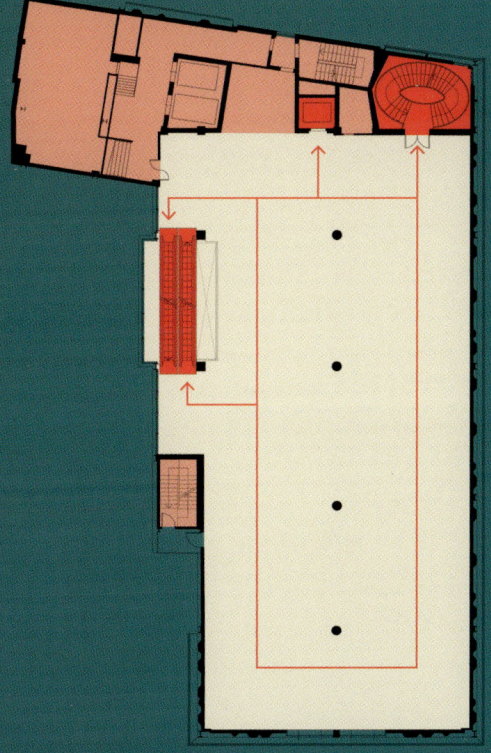
TYPE PLAN / STATE OF AFFAIRS

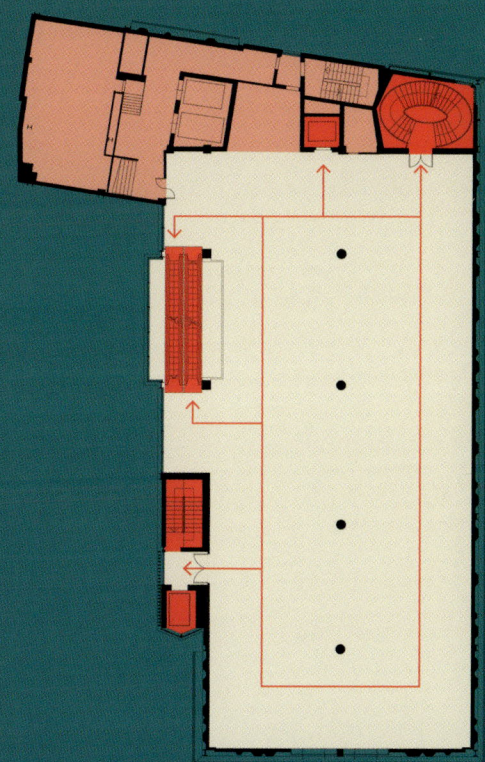
6TH FLOOR PLAN

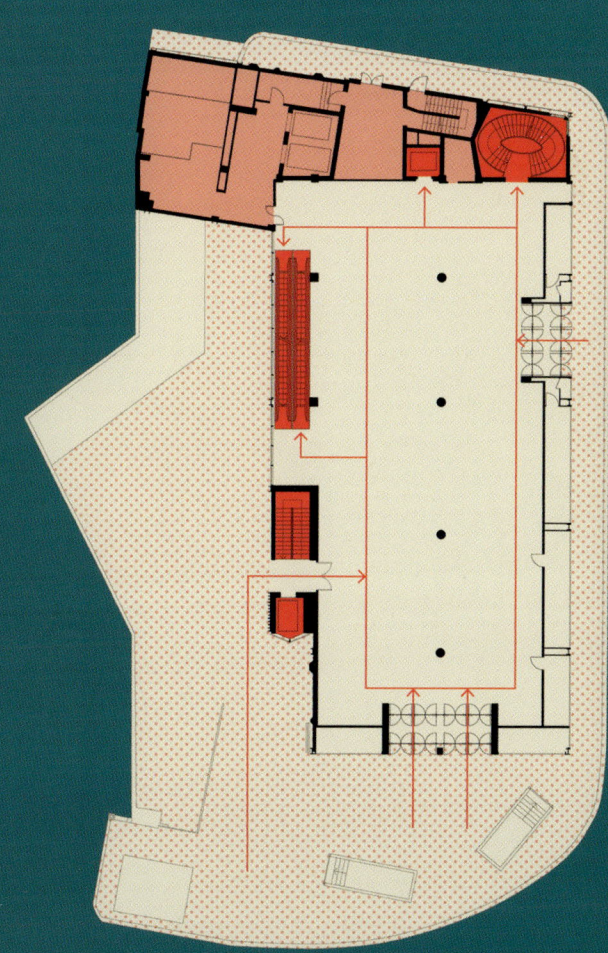

circulation
vertical circulation
sales space
back-of-house

GROUND FLOOR PLAN

NORTH ELEVATION

WEST ELEVATION

엔하코 빌딩
ENHAKO BUILDING

ARCHITECT : ORGANIC DESIGN INC. / HIDEO KUMAKI

THE ORIGIN OF "EN-HAKO" MEANING CONSISTS OF TWO ELEMENTS. The first "En" means "Engawa" which has traditional roots as a garden seat for welcoming guests associated with Japanese-style dwellings from medieval times in Japan. The rear "hako" means the box or the venue, so to speak, refers to the building itself, which contains the functional tenant space for the facility.

The En-haco Building is located between residential neighborhoods and a commercial/government area in the suburbs of a commuter town, for a while walking from the train station, where older residents, homemakers, and children come and go during the week for various purposes. At this boundary where such urban attractions await, we sought to provide a perch-literally, on verandas where people can relax for a moment—and a familiar landmark for the community. It was insightful to consider people's movement at this place where people are coming and going both during the week and on weekends.

The surroundings do not allow for the kind of foot traffic at train stations or other crowded sites, but there is certainly an ebb and flow from the community. Thus, the building itself has a roadside veranda covered by balconies that are like a pair of elevated, undulating verandas themselves. From below, these form wooden eaves with enticing, organic shapes that draw people nearer to the building. It faces north, providing a vantage on park greenery, city hall, and other sights across the boulevard. The atmosphere of summer festivals, plant fairs, and other lively local events is tangible from here.

Zoned for medical and retail offices, the Enhaco Building with its inviting veranda was designed as if it were reaching out to the neighborhood, beckoning one and all to stop by on their way between residential and commercial areas through the week. Functionality and infrastructure are required within the facility, which has been designed to offer a choice between renting the entire building, specific floors, or specific units.

'엔하코(En-hako)'의 어원에는 두 가지 의미가 담겨있다. '엔'은 일본 중세 주택에서 정원을 바라보는 좌석 공간으로 사용된 '엔가와'라는 일본 전통에서 기원을 두고 있다. 박스 또는 상자를 의미하는 '하코'는 말하자면 기능적인 세입자 공간을 포함한 건물 자체를 의미한다.

엔하코 빌딩은 도심으로 통근하는 사람이 많은 교외 주거 지역과 상업 지구 및 관공서 사이에 위치하고 있으며, 기차역에서 도보로 이동이 가능하여 주중에 남녀노소를 불문하고 다양한 이동인구가 모인다. 도시의 경계 지역에 위치한 지역적 특성을 고려해 이 건물은 말 그대로 사람들이 어느 때나 편하게 와서 쉴 수 있는 베란다 공간이자 커뮤니티의 랜드마크가 될 휴식 공간을 제공한다. 주중과 주말 이동 인구가 많은 이곳에서는 사람들의 동선을 고려하는 것이 요구되었다.

주변을 둘러보면 기차역이나 사람들이 붐비는 곳에서 사람들의 동선이 복잡하지는 않지만, 일정한 흐름이 관찰되었다. 이에 길가에 면한 건물 위아래에 물결 모양을 이루는 발코니로 둘러싸인 베란다 공간을 지면에서 띄워 조성하였다. 이는 자연스러운 형태의 나무 처마를 연상시키며, 아래에서 바라보는 사람들을 건물로 끌어들이는 디자인 요소이다. 녹지 공원, 시청 및 대로 건너편을 조망할 수 있는 북향 건물로, 여름 축제, 식물 박람회 및 여러 지역 행사의 활기찬 분위기를 느낄 수 있다.

의료 및 소매상점 구역으로 나뉜 엔하코 빌딩에 조성한 베란다는 일주일 내내 주거 지역과 상업 지구를 오가는 사람들에게 손을 뻗어 초청하는 듯한 분위기를 연출한다. 건물 전체, 특정 층 또는 특정 단위를 임대할 수 있도록 설계된 시설 내에 기능성 및 인프라 구축이 필요하였다.

Location Saitama, Japan **Use** Office **Site area** 492.27m² **Building area** 283.89m² **Gross floor area** 784.86m² **Completion** 2023 **Project manager** Hideo Kumaki, Rei Maki **Contractor** GUNPOH INDUSTRIES Co.,Ltd. **Photographer** Yukinori Okamura

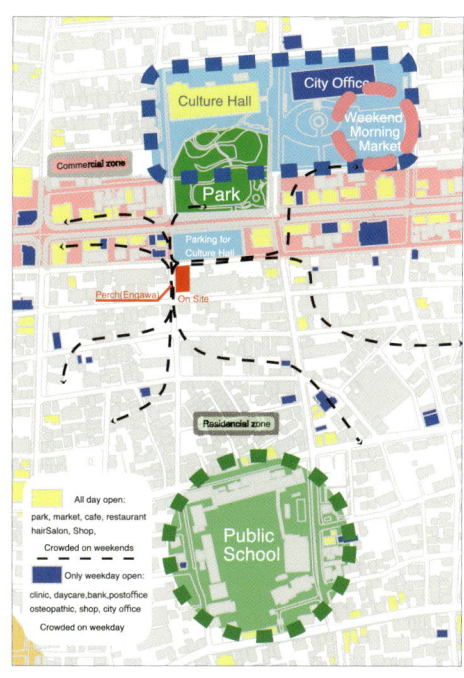

THE PERCH OF CITY

← Corner view

↑ View from west　← Looking-up view　→ Detail view of veranda　↙ Building signboard　↳ Detail view of Balcony

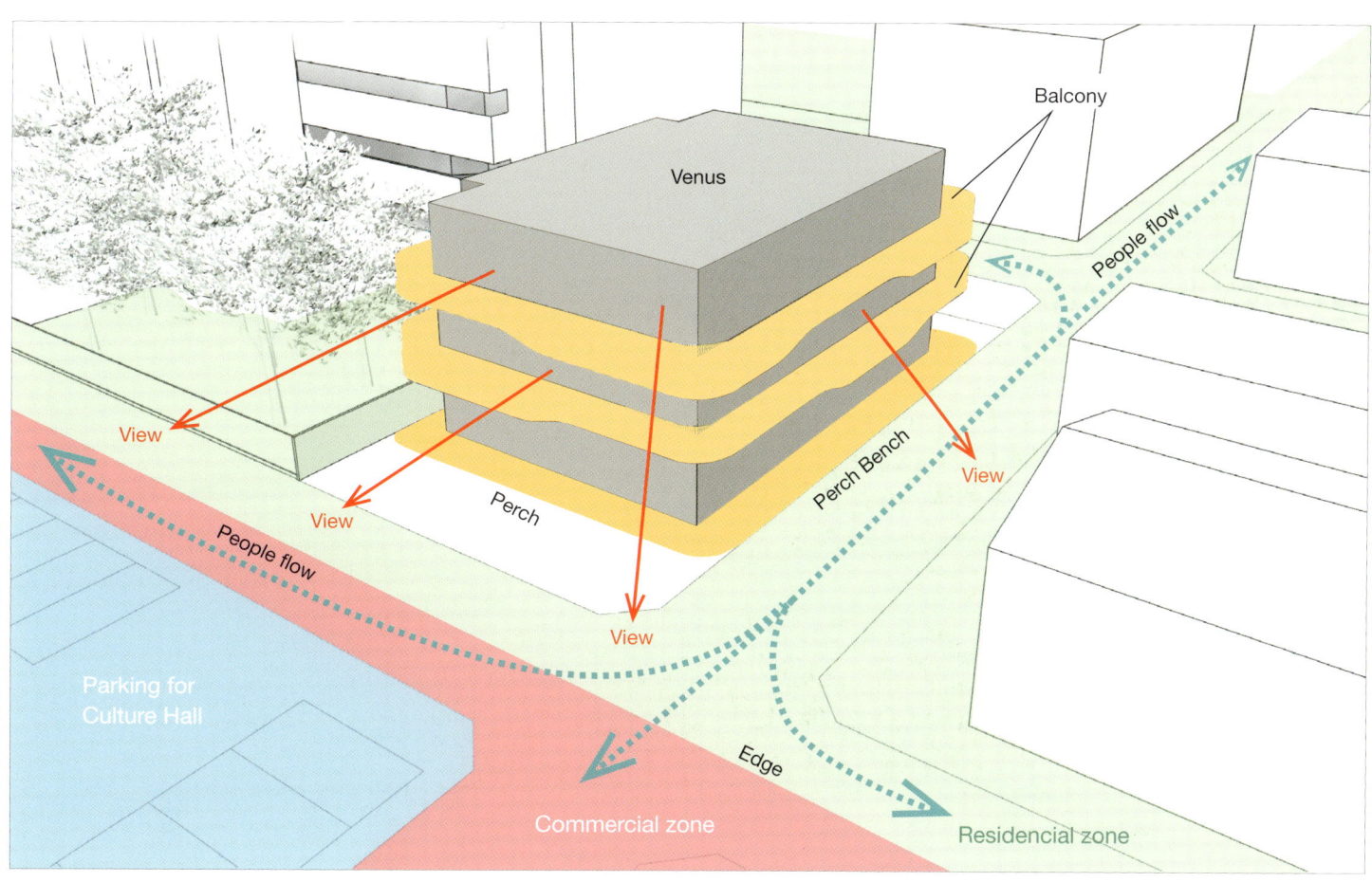

↑ Exterior view

BUILDING COMPOSITION

↑ Corner view ← Exterior view → Exterior view

LONGITUDINAL SECTION CROSS SECTION

← Veranda   → Looking-up view

1 SHEAT WATERPROOFING 110 SPEC. + 12T STRUCTURAL VENEERS 2P
2 C-CHANNNELS-100X50X20X2.3 @303
3 H-STEEL-200X100
4 C-CHANNNELS-50X50 @303
5 FURRING STRIPS-50X50 @303 HOT DIP GALVANIZING
6 Ø60 WATER PIPES
7 HANDRAIL : STAINLESS STEEL + T6 HOT DIP GALVANIZING VERTICAL BARS @80 TOP PLATE FLAT BAR+TABLE TOP
8 STEEL L-TYPE ANGLE BRACKET : L6×50 WELDING PARTS + RUST PROOF PAINT
9 BRIDGE BOARD SPEC : T9 STEEL HOT DIP GALVANIZING
10 LOWER PARTS OF BALCONY : T15 WOODEN BOARD + OIL STAIN PAINTS
11 GUTTER WATER SLOPE 1/100
12 TABLETOP : DECKING MATERIAL : 139X22X2000 MODERN BLACK GAPPED BOARDS/ART WOOD
13 T5 STAINLESS STEEL FLAT BAR

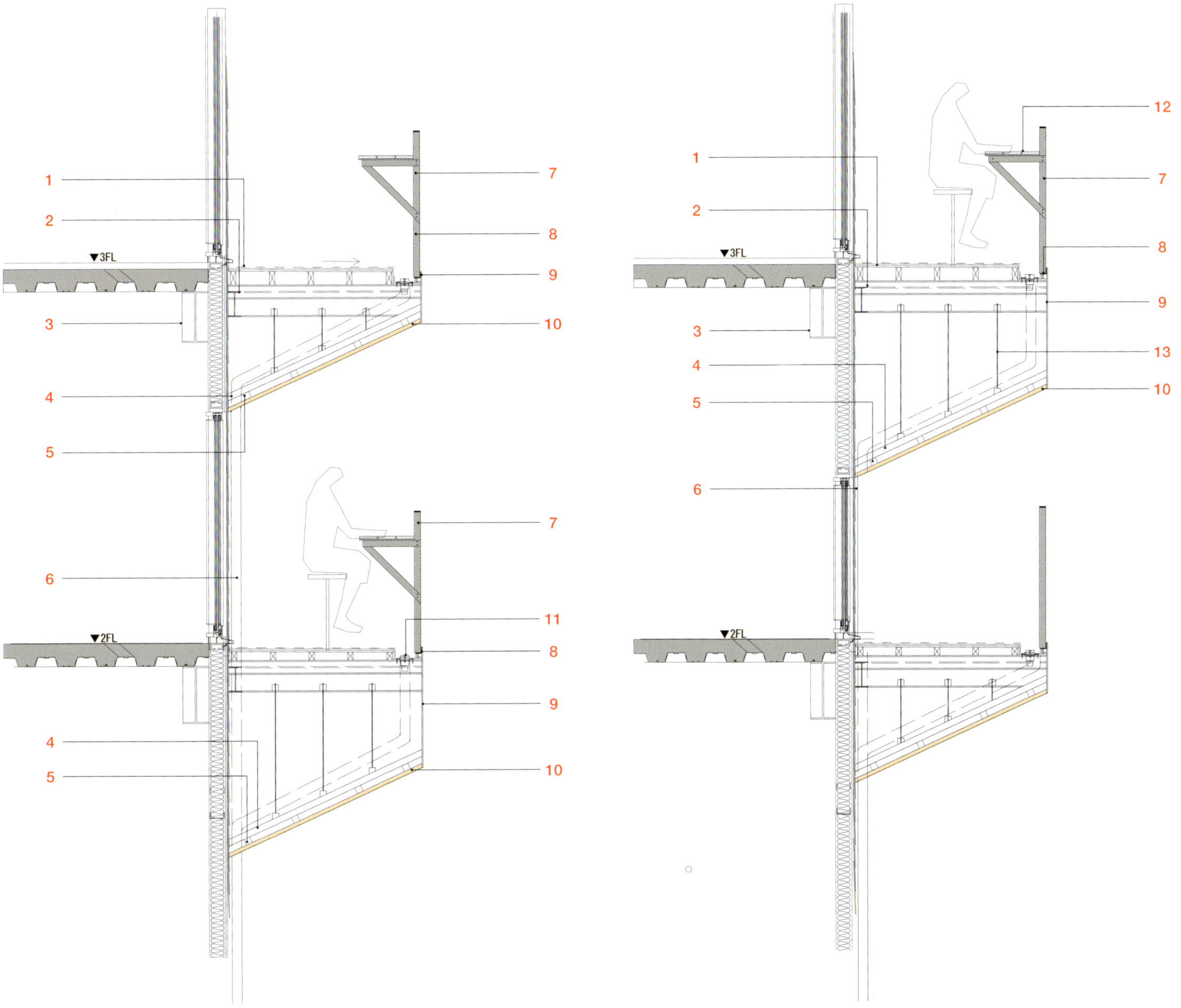

SECTION DETAIL

↗ Interior view of the 3rd floor
← Interior view of the 2nd floor
↙ Interior view of the 1st floor

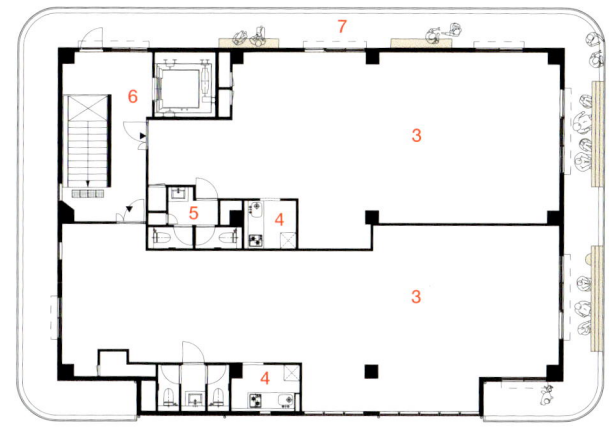

↑ Interior view of the 1st floor

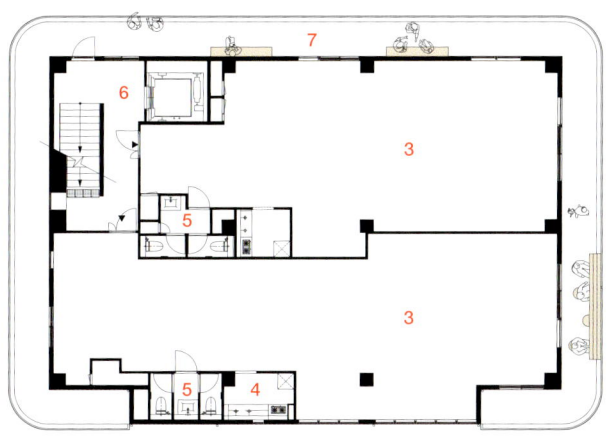

2ND FLOOR PLAN

3RD FLOOR PLAN

1. PARKING LOT
2. TERRACE
3. ROOM
4. KITCHENETTE
5. RESTROOM
6. CORRIDOR
7. VERANDA

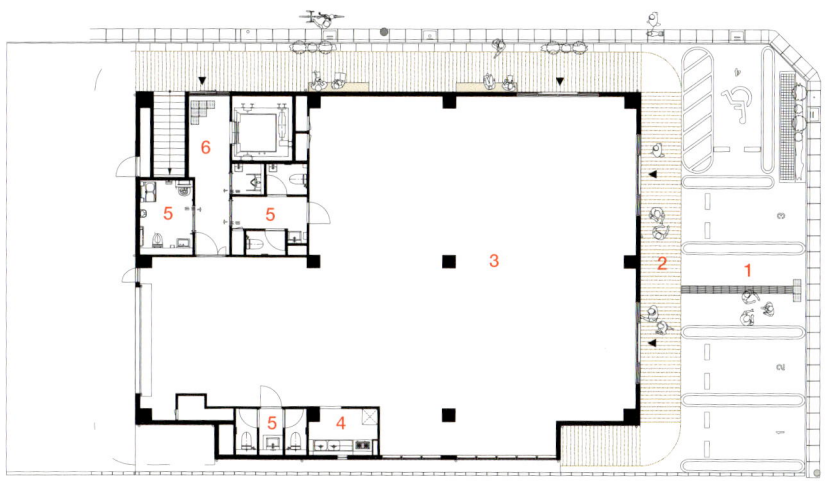

1ST FLOOR PLAN

↑ Staircase   ↖ Corridor   ↙ Elevator hall   → Corridor

↑ Veranda

1. DECKING MATERIAL : 139X22X2000 MODERN BLACK GAPPED BOARDS/ART WOOD
2. T6 427×75 STAINLESS STEEL FLAT BAR +HOT DIP GALVANIZING FILLET WELDING
3. T6 450×50 STAINLESS STEEL FLAT BAR + HOT DIP GALVANIZING FILLET WELDING
4. T6 50 STAINLESS STEEL FLAT BAR + HOT DIP GALVANIZING FILLET WELDING
5. STAINLESS STEEL + T6 HOT DIP GALVANIZING VERTICAL BARS @80 FILLET WELDING
6. STEEL L-TYPE ANGLE BRACKET : L6X50 FILLET WELDING + RUST PROOF PAINT
7. BRIDGE BOARD SPEC : T9 STEEL HOT DIP GALVANIZING
8. SEALING AGENT
9. DECKING MATERIAL : 139X22X2000 MODERN BLACK GAPPED BOARDS/ART WOOD
10. T9 STAINLESS STEEL FLAT BAR + HOT DIP GALVANIZING FILLET WELDING
11. T6 400X75 STAINLESS STEEL FLAT BAR + HOT DIP GALVANIZING FILLET WELDING
12. CAST IRON FUNNEL DRAIN
13. SHEAT WATERPROOFING 110 SPEC. + 12T STRUCTURAL VENEERS 2P

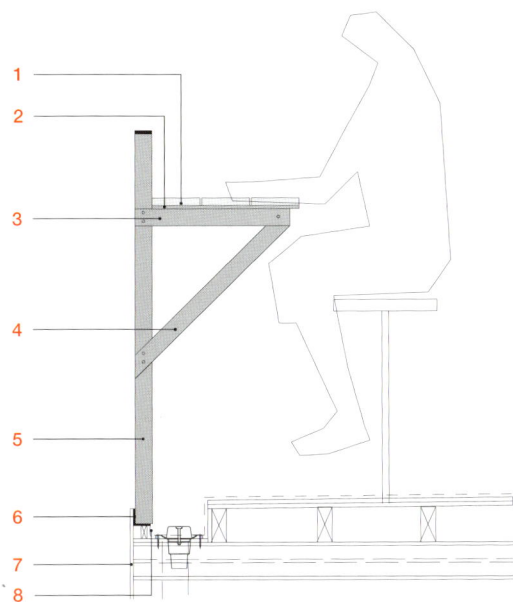
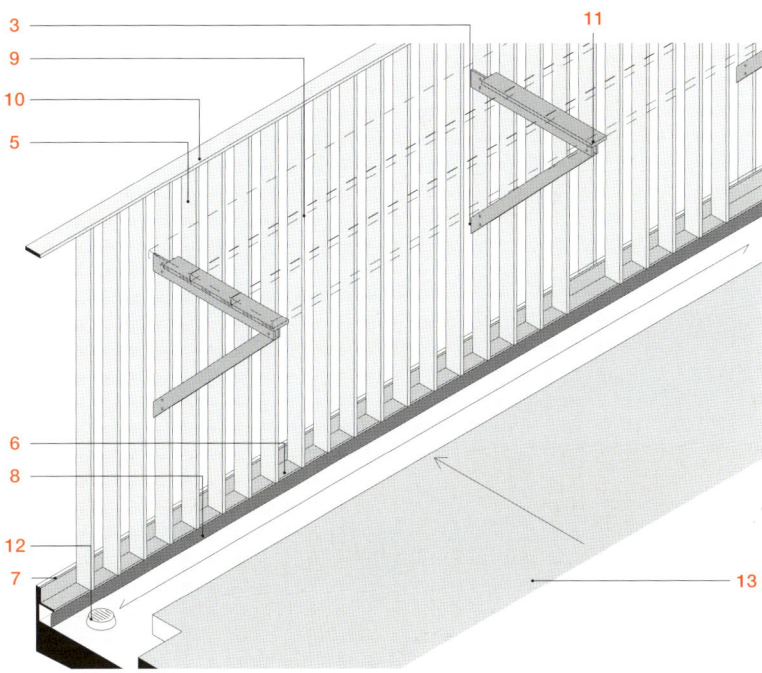

SECTION DETAIL

도시 정원

URBAN GARDEN

ARCHITECT : LOIZOU ARCHITECTS + ASSOCIATES / GEORGE LOIZOU

THE NOTION OF PLANTS, TREES AND PEOPLE COEXISTING UNDER ONE ROOF can be characterized timelier and more necessary than ever. While the horizontal ground of our cities is being covered by the construction front, the vertical ground is evolving, creating living buildings which are deemed viable cells in the urban fabric and contribute to improving the quality of life of residents and neighboring schemes.

The building is located in the densely populated area of Ayios Nikolaos, east of the centre of Limassol. An area characterized by the single-family houses of the 60s and 70s and that has been undergoing intense urban and architectural changes in the last decade.

The building is divided into five levels, the piloti, three storeys and the roof garden. The piloti allows access to the building and is also the parking area. The first-floor houses an architecture studio space with a drawing room and a meeting room. The second and third floors are floor apartments which include an open plan kitchen, living room, dining room, two bedrooms and two bathrooms. The open plan apartments are surrounded by south and west facing openings while the bedrooms are west and north facing. All interiors are surrounded by sliding thermal aluminum glazing. On the one hand, they help reduce energy loss while on the other hand they unify the exterior and interior spaces, allowing the user to have an unobstructed view of the green facade. The roof garden consists of a dining area, lounge area and barbecue facilities. A metal pergola covers the entire rooftop. Double-sided photovoltaic panels are located on the upper part of the pergola, helping to reduce the energy consumption of the building and creates shade during the daytime.

Location 10 Mahatma Gandhi Str. Agios Nikolaos, 3096 Limassol, Cyprus **Use** Residential Building with Office **Site area** 300m² **Gross floor area** 360m² **Completion** 2022 **Contractor** Medusa Constructions **Photographer** Maria Efthymiou

식물과 나무, 사람이 한 지붕 아래 공존한다는 개념은 그 어느 때보다 시의적절하며 절실하다. 현대 도시의 수평적 지면 위로 건축물이 빽빽하게 세워지는 동안 수직적 지면은 진화하여 마치 살아 있는 듯한 도시 조직의 세포로 변모함으로써 주변 지역에 살고 있는 사람들의 삶의 질을 향상하는 데 기여하게 되었다.

리마솔 중심부 동쪽에 위치하고 인구 밀도가 높은 지역인 아요스 니콜라오스에 세워진 건물을 소개하고자 한다. 이곳은 60~70년대 단독주택이 밀집한 지역으로 지난 10년간 도시와 건축 디자인이 급격하게 변화하였다. 총 5층으로 이루어진 이 건물은 필로티와 3개 층의 공간 및 옥상 정원으로 구성되어 있다. 필로티는 건물에 접근할 수 있는 곳으로, 주차장으로도 사용된다. 1층에는 응접실과 회의실을 갖춘 건축 스튜디오가 자리한다. 2층과 3층은 개방형 주방과 거실, 식당, 침실 2개, 욕실 2개를 갖춘 아파트입니다. 개방형 구조의 아파트는 서쪽과 남쪽이 통창으로 둘러져 있고, 침실은 서쪽과 북쪽을 향하고 있다. 내부 전체는 온도 조절 기능을 위해 슬라이딩 알루미늄 창으로 마감하였다. 이는 열 손실을 줄이는 동시에 내부와 외부의 공간을 통합하여 사용자가 창밖의 녹색 식물을 보는 것에 방해받지 않을 수 있다. 옥상 정원은 라운지 공간과 바비큐가 가능한 식사 공간으로 구성되어 있습니다. 옥상 전체에는 금속 퍼걸러를 구성하고 퍼걸러 상부에는 이중 태양광 패널을 더해 건물의 에너지 소비를 줄이고 낮에는 자외선을 차단해준다.

← Front facade   → Partly view of exterior

↑ Exterior view

SOUTH ELEVATION

EAST ELEVATION

NORTH ELEVATION

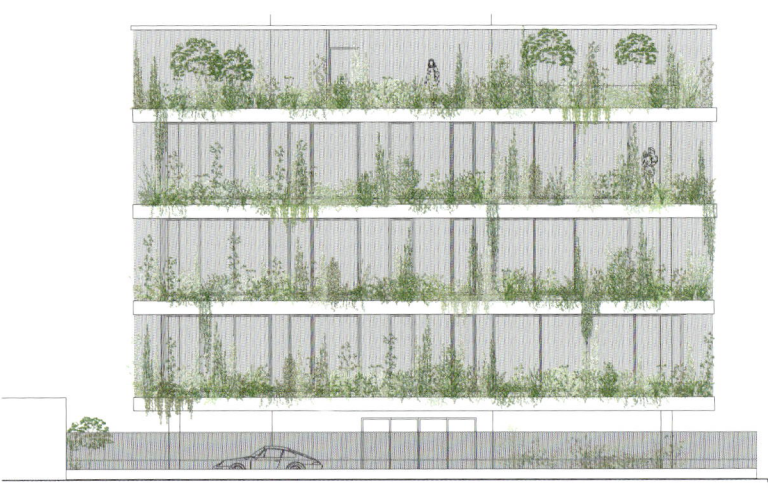

WEST ELEVATION

← Night view　→ Night view　↳ View from west

↑ Panoramic view

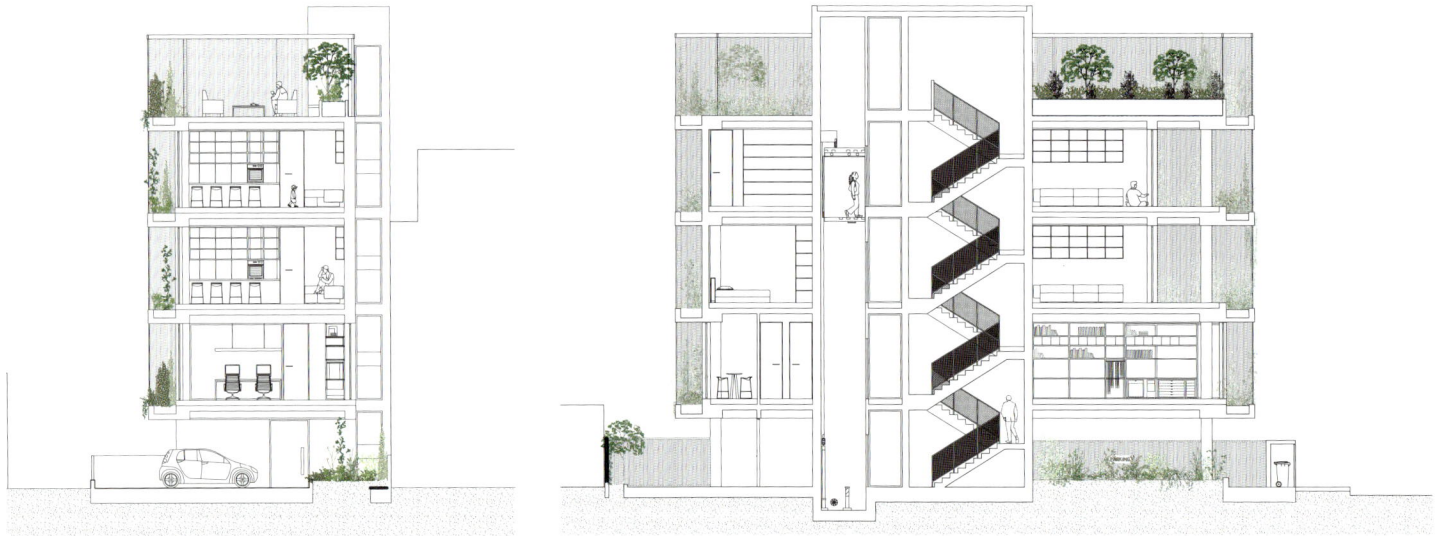

CROSS SECTION **LONGITUDINAL SECTION**

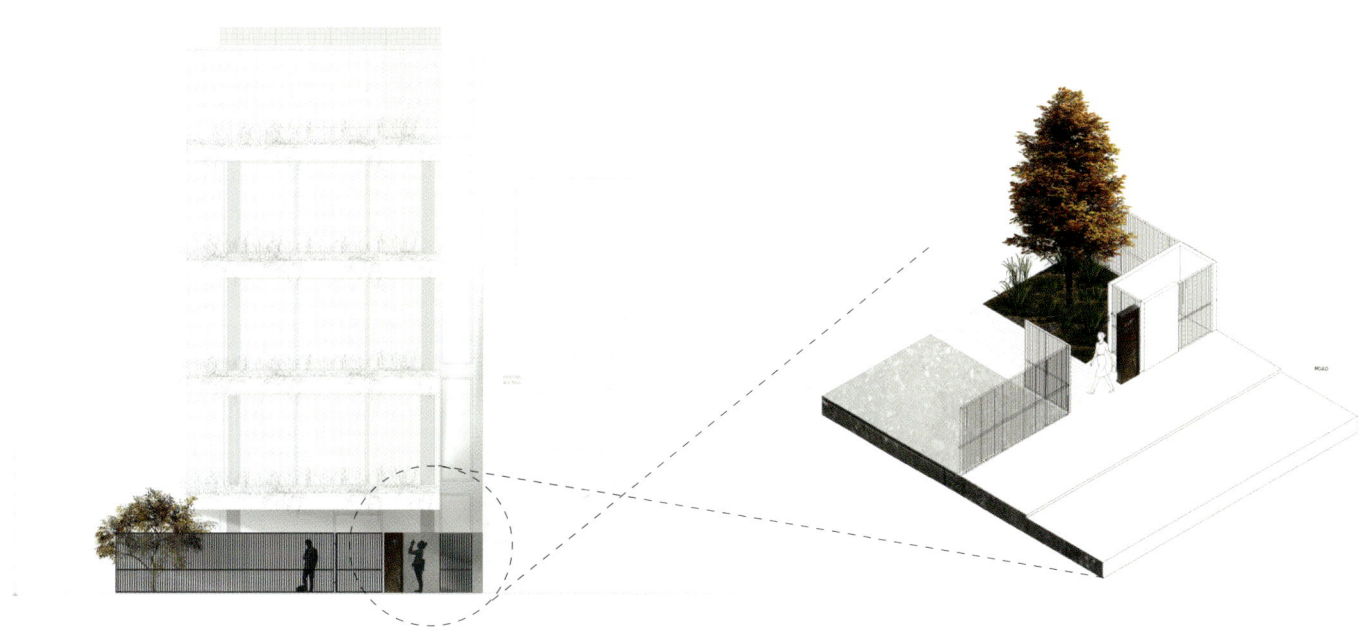

↑ Main entrance

FENCE ELEVATION

- Office
- Corridor
- Partly view of exterior

↑ Office

1 PARKING LOT
2 ENTRANCE
3 MEETING ROOM
4 ELEVATOR HALL
5 OFFICE
6 REST TABLE
7 RESTROOM

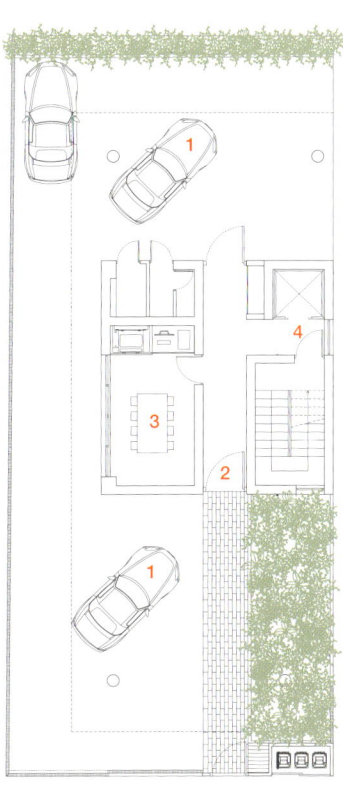

GROUND FLOOR PLAN

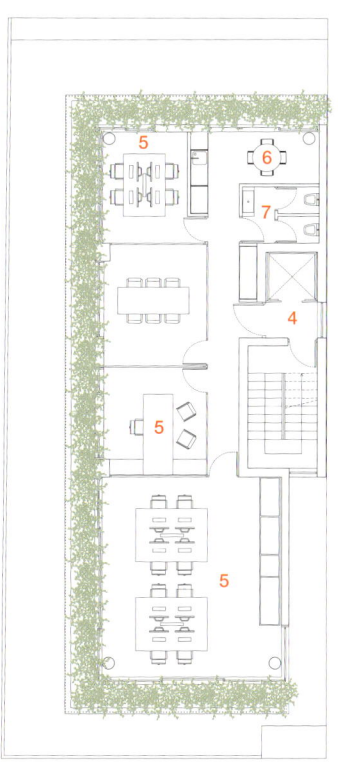

1ST FLOOR PLAN

↑ Residential unit

1 LIVING & DINING ROOM	4 BEDROOM	7 ELEVATOR HALL	10 RESTROOM
2 KITCHEN	5 MASTER BEDROOM	8 OUTDOOR KITCHEN	11 SOLAR PANEL
3 BATHROOM	6 MASTER BATHROOM	9 ROOFTOP GARDEN	12 FLOLWER BED

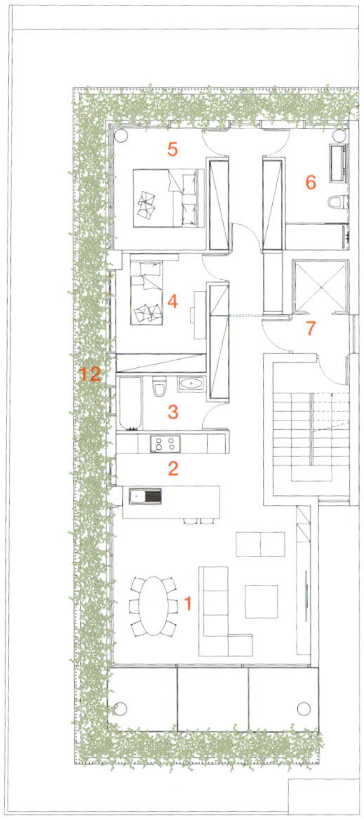

2ND FLOOR PLAN 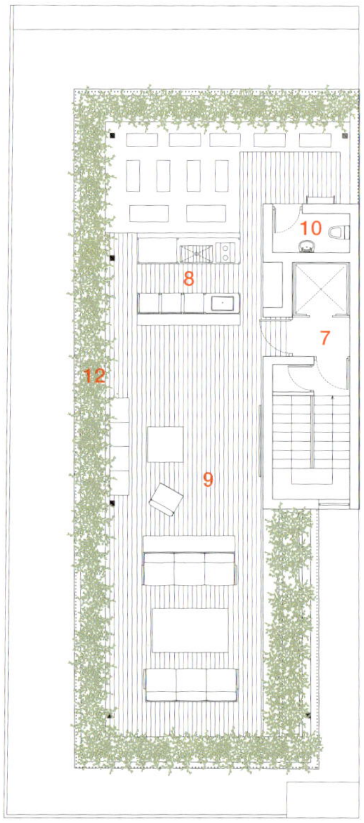   **ROOF FLOOR PLAN** 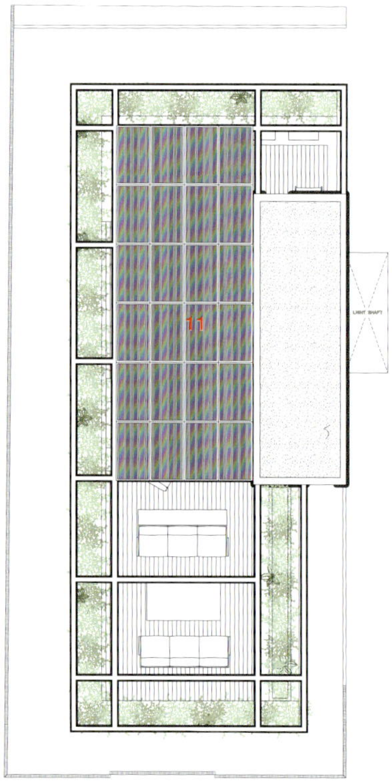   **ROOF PLAN**

← Interior view   → Rooftop garden

1 ESTAINLESS STEEL WIRE ROPE - 6mm DIAMETER
2 REINFORCED CONCRETE
3 WEDGE ANCHOR
4 PUSH - LOCK STUD
5 STAINLESS STEEL PLATE
6 HEXAGOLAN STAINLESS STEEL NUT
7 HEXAGONAL STAINLESS STEEL DOME BOLT
8 ELECTRIC WELDING
9 IPE BEAM 240
10 GALVANISED FLASHING
11 SOIL
12 GEOFABRIC
13 GRAVEL
14 WATER INSULATION
15 SCREED
16 REINFORCED CONCRETE
17 ALUMINIUM SLIDING DOOR
18 SCREED
19 WINDOW SILL
20 BRICK

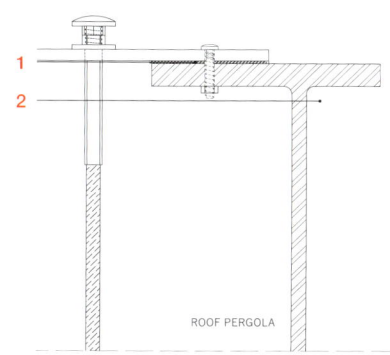
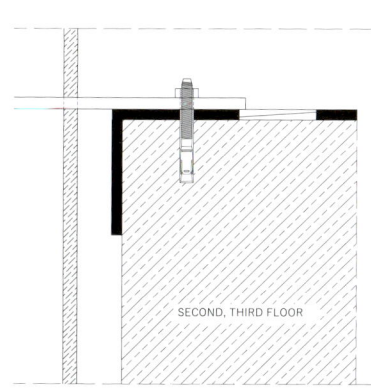
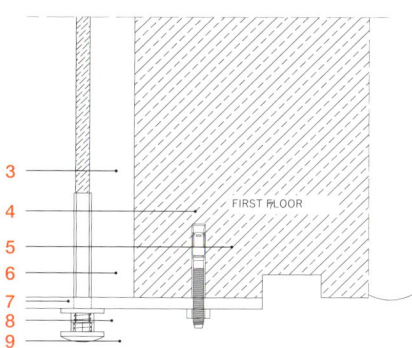

SECTION DETAIL   SECTION DETAIL

LOCATED IN A DENSE NEIGHBORHOOD IN FRONT OF a commuter train station, the client is a foundation that wants to build 3 stories mosque with a school on the ground floor serves as a verandah for the surrounding community as a replacement from its previous location.

A timeless architecture : Our approach is to design a timeless mosque by having a strong approach to the site context such as climate, culture, people's habits, and passive design strategies. We believe that the architectural form of the mosque can develop and be unique in different contexts because it embodies the identity of the community it shelters and a power to become the genius loci of the surrounding neighborhood.

Architectural Strategies : We design architecture that is not bound by time but has a strong environmental context by building a dialogue with the sun, the climate, and high intensity of public interaction in this location that appears in the strategy and architectural design elements as our 'translation'.

Three sloping roofs creating sequences : Each roof segment divides the area of the mosque into a welcoming porch, a terraced worship area, and a main worship area. The sloping roof provides shade and responds to rain and heat. At the tip of each roof, there is a 'skylight' that establishes a dialogue between worshipers and the time; through which the natural light shows different praying times of the day.

Symbolizing the word 'Allah' on the integration of the roof and minaret shapes : The integration of the 3-roof segments and the tower forms a symbol of the word Allah when viewed from the side of the direction of the train terminal.

Thermal Comfort : This mosque is designed without artificial air conditioning by providing optimum cross ventilation, the synthetic wood lattice appears to provide a curtain and control overheat penetration that enables the building to maintain a comfortable thermal comfort.

Stairs, terraces, and courtyards as public spaces : The function of the mosque serves not only for religious purposes but also as a center for community interaction that is presented through a courtyard without a fence that invites activities from the surrounding community as well as stairs on the perimeter of the building as an element to sit and an accessible public space.

Location Jl. Batu Merah IV Blok Rv 22 No.42, RT.5/RW.2, East Pejaten, Pasar Minggu, South Jakarta City, Jakarta 12510, Indonesia **Use** Mosque **Site area** 1.260,32m² **Built area** 360,39m² **Gross Floor area** 745m² **Completion** 2020 **Design team** Muhammad Luthfan Rizal, Danindra A. Wicaksono **Contractor** Pt. Prodecon Mitratama **Photographer** Kafin Noe'man

자미알 후리야 빛의 사원
MASJID CAHAYA JAMI AL-HURRIYAH
ARCHITECT : AGO ARCHITECTS / ABIMANTRA PRADHANA, OSRITHALITA GABRIELA

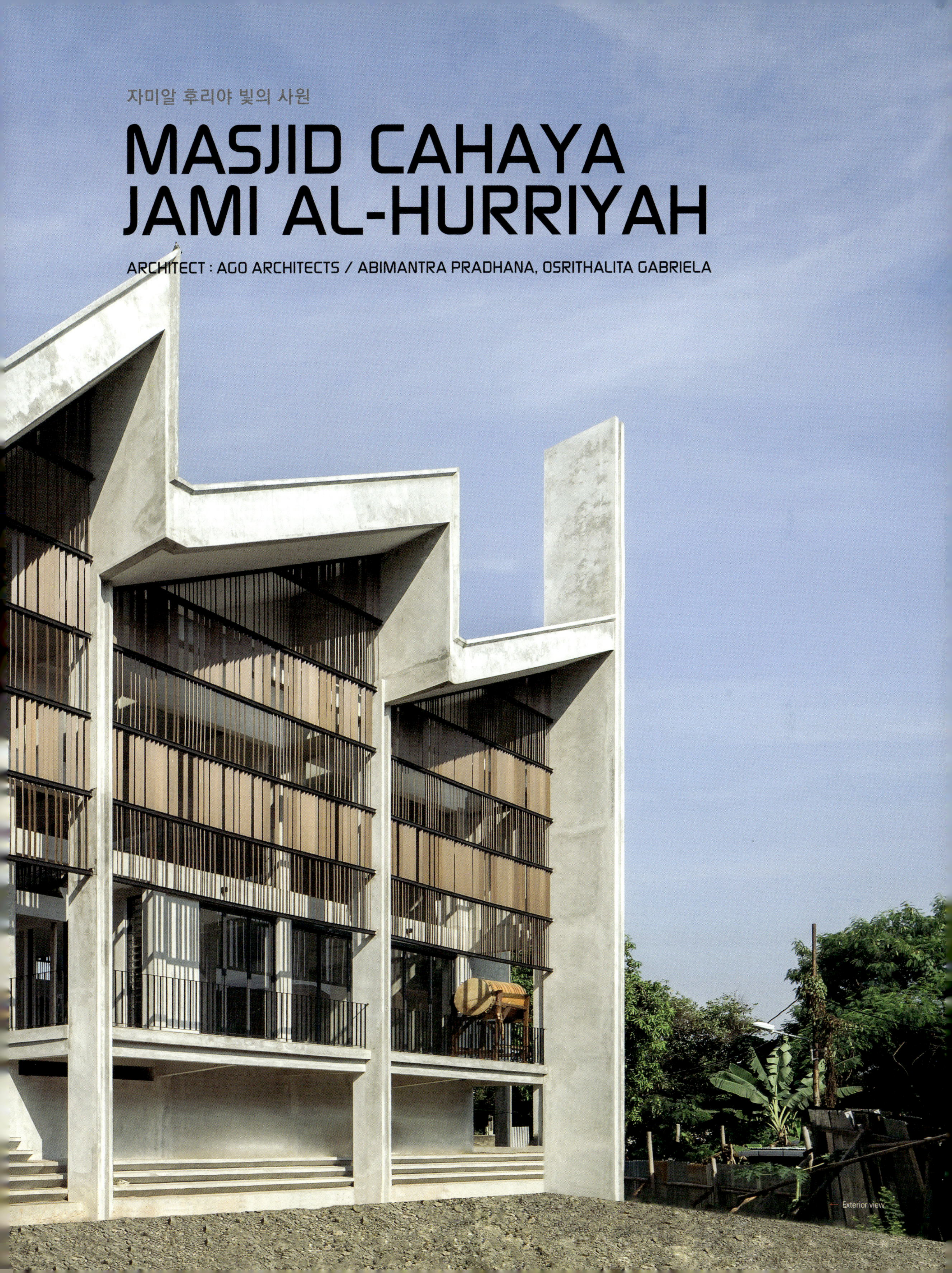

Exterior view

← Stairs as public spaces

통근 기차역 앞 밀집된 동네에 위치한 건물의 소유주는 재단으로, 기존 건물을 3층 규모의 모스크 건물로 재건축하고 1층에서는 학교를 운영하며 주변 커뮤니티의 교두보 역할을 원했다.

시간이 흘러도 변함없는 건축물 : 우리는 기후, 문화, 사람들의 습관 및 수동적 디자인 전략과 같은 사이트 컨텍스트에 전적으로 따라 시간이 흘러도 변함없는 모스크를 설계하고자 했다. 모스크의 건축 형태는 보호하려는 공동체의 정체성과 힘을 구성해 그 주위로 전능한 힘을 발휘하는 장소라는 맥락에서 독특하게 발전시킬 수 있다고 믿었다.

건축 전략 : 시간의 흐름에 메이지 않고 해당 지역의 빛, 기후, 공공 상호 작용의 밀집도에 따른 환경적 맥락을 디자인 요소로 '번역' 하는 건축 전략을 세웠다.

세 개의 경사진 지붕이 연결된 구조 : 각 지붕은 환영이 이루어지는 현관, 예배가 진행되는 테라스 및 주요 예배실로 이루어진 모스크의 각 공간을 분리했다. 경사진 지붕은 직사광선, 열 및 빗물을 막아준다. 각 지붕의 끝에는 시간의 흐름에 따라 예배하는 사람을 위한 '채광창' 이 있어 자연 채광을 통해 기도 시간을 알 수 있다.

지붕과 연결된 미나레트에 깃든 '알라' 의 상징성 : 세 개의 지붕과 탑을 연결해 기차 터미널 방향에서 볼 때 알라라는 상징성을 표현하고 있다.

온도 조절 기능 : 인공 에어컨 없이 최적의 교차 환기를 제공하도록 설계된 모스크에 합성 목재로 격자를 설치해 직사광선을 차단하여 건물 내부의 온도를 조절하고 쾌적하게 유지할 수 있다.

공공 공간으로서의 계단, 테라스 및 안뜰 : 종교적 목적뿐 아니라 주변 커뮤니티와 상호 작용과 모스크의 기능을 강화하기 위해 초청인들과 액티비티를 진행할 수 있도록 안뜰에 벽을 없애고, 방문객이 쉽게 접근할 수 있도록 건물 주변에 공공장소로서의 계단을 구성했다.

↑ Entrance & stairs

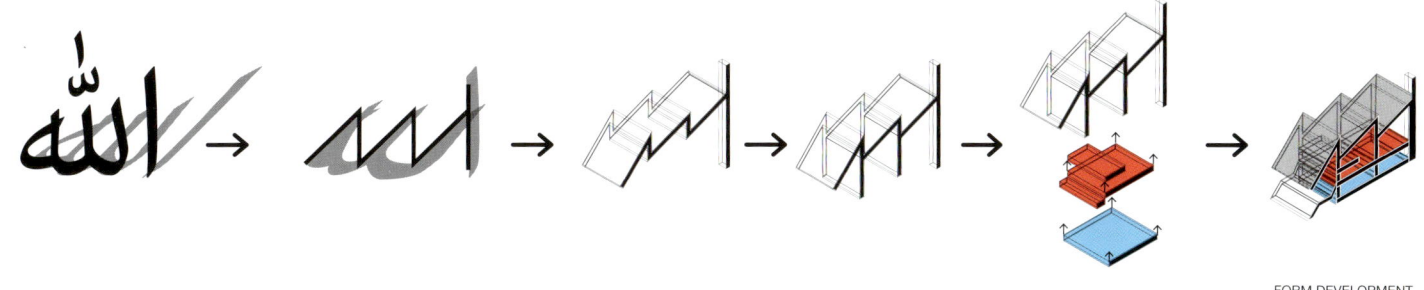

FORM DEVELOPMENT

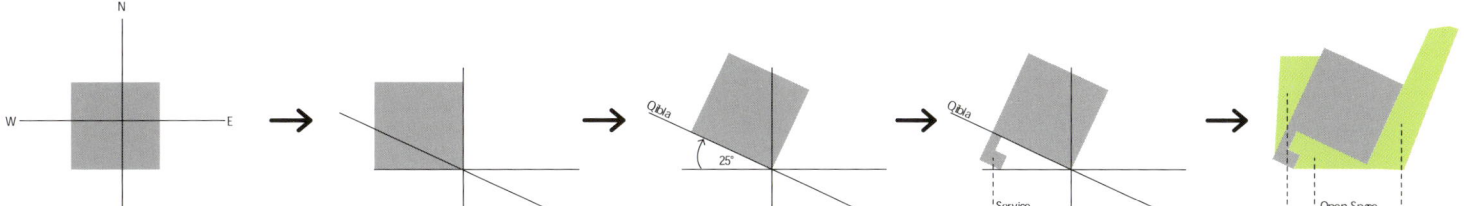

ORIENTATION DEVELOPMENT

DIAGRAM

↑ Sloping roofs   ↙ Terraced worship area   ↓ Terraced worship area   ↘ Terrace, the synthetic wood lattice

↑ 3 stories mosque, veranda
→ 3 stories mosque, veranda
↳ 3 stories mosque, veranda

↑ Facade, the synthetic wood lattice

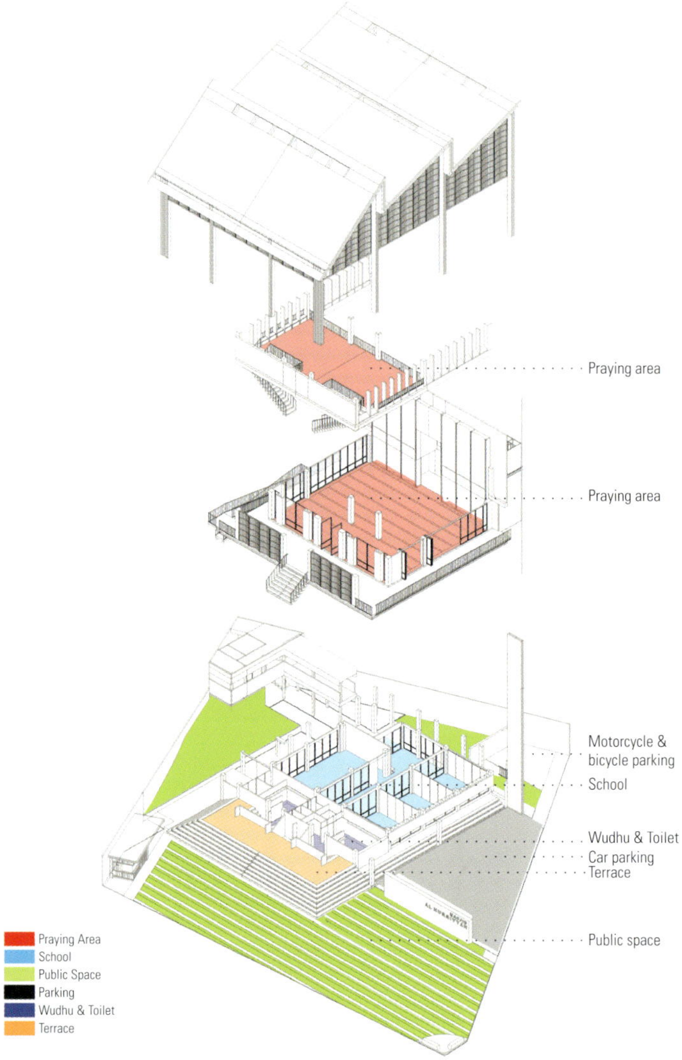

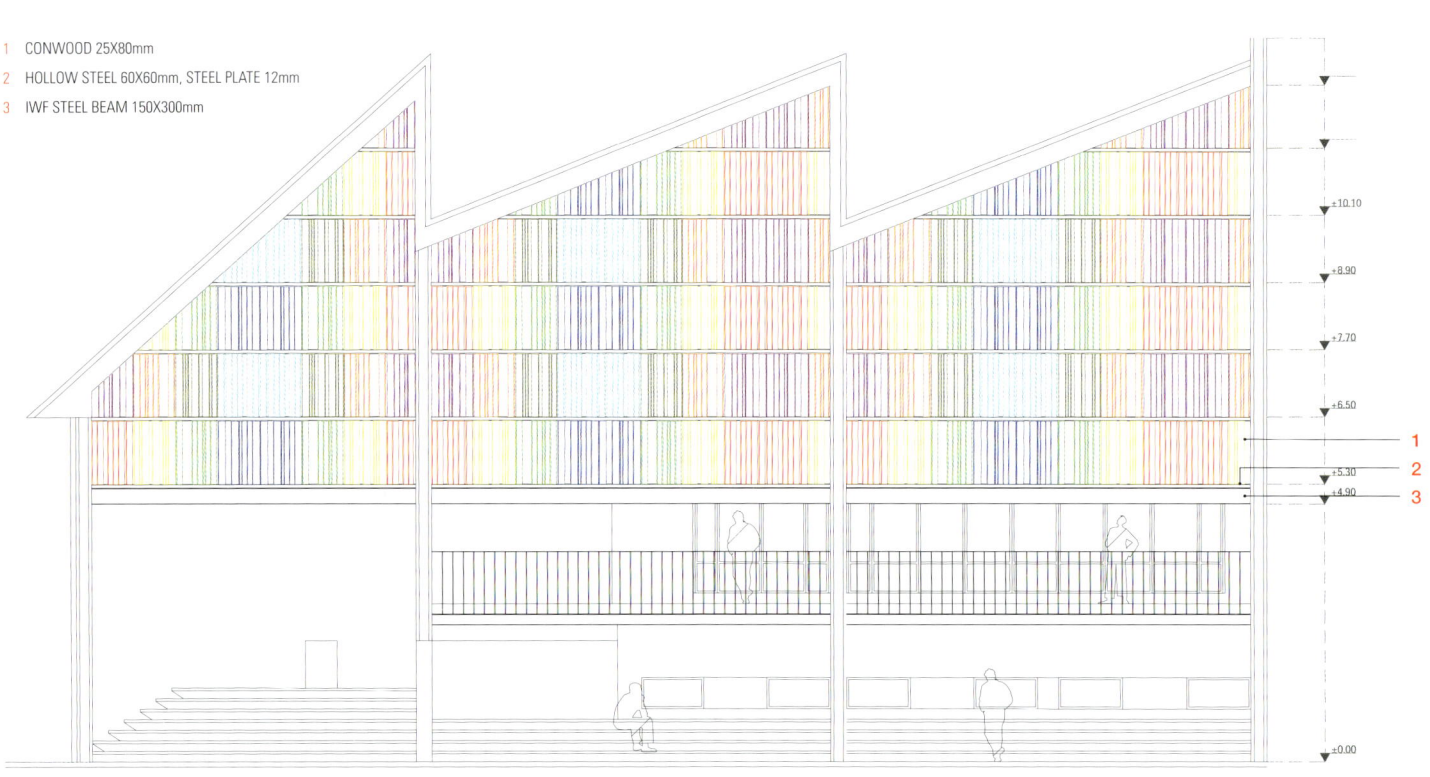

ISOMETRIC

1. CONWOOD 25X80mm
2. HOLLOW STEEL 60X60mm, STEEL PLATE 12mm
3. IWF STEEL BEAM 150X300mm

ELEVATION

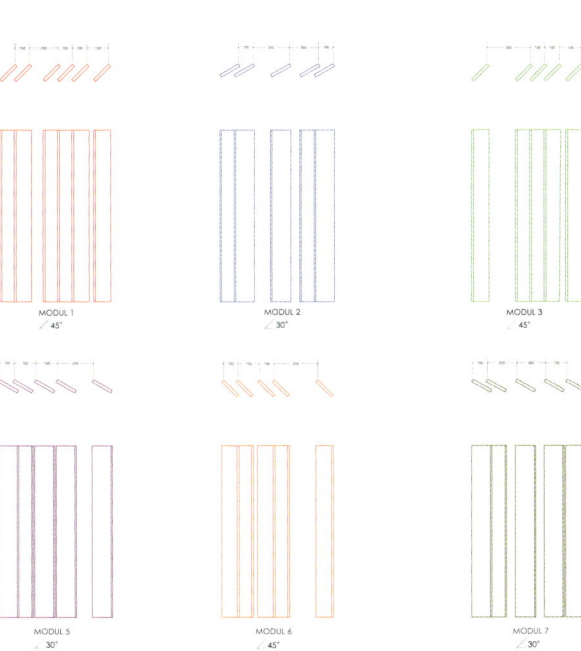

FACADE MODULE DETAIL

↑ Terrace, the synthetic wood lattice

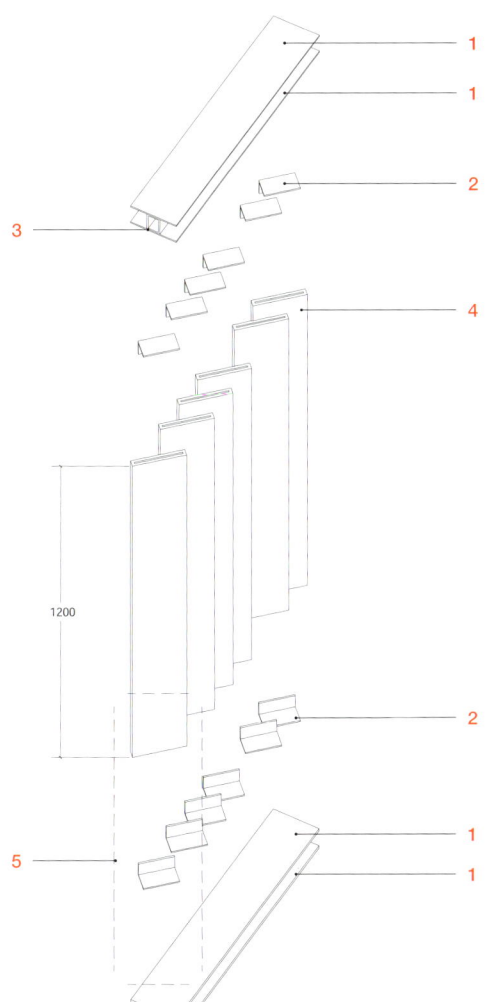

1 STEEL PLATE WIDTH 150mm, THICKNESS 5mm
2 L ANGEL STEEL 50mm, THICKNESS 5mm
3 HOLLOW STEEL 60X60X2mm
4 CONWOOD 25X80mm
5 DETAIL RERERENCE
6 BOLT

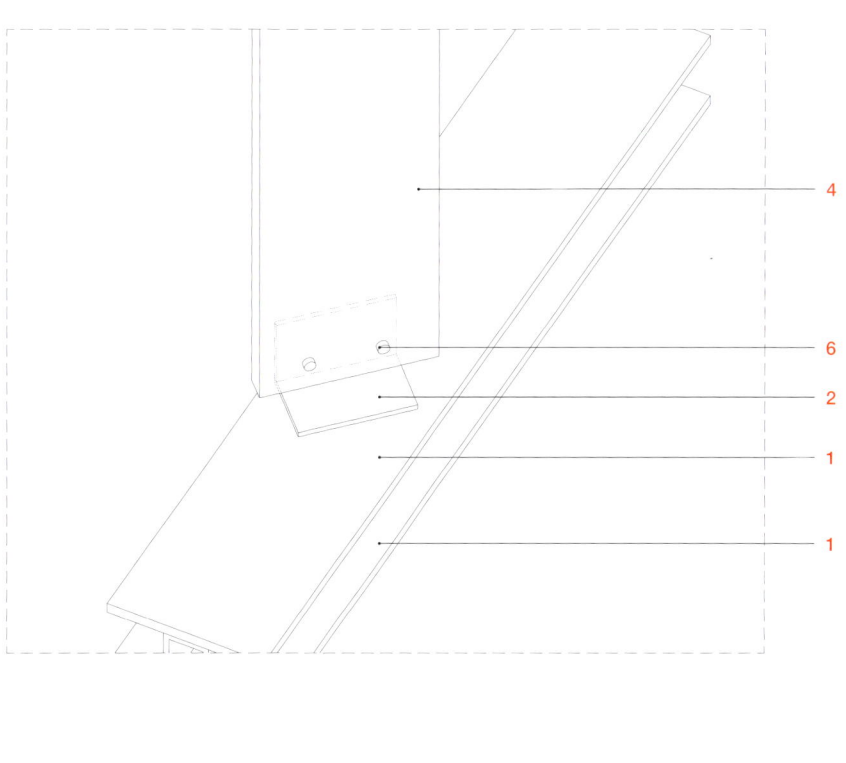

JOINT DETAIL

↖ Facade, the synthetic wood lattice   ↗ Inside, main praying area   ↓ Praying area

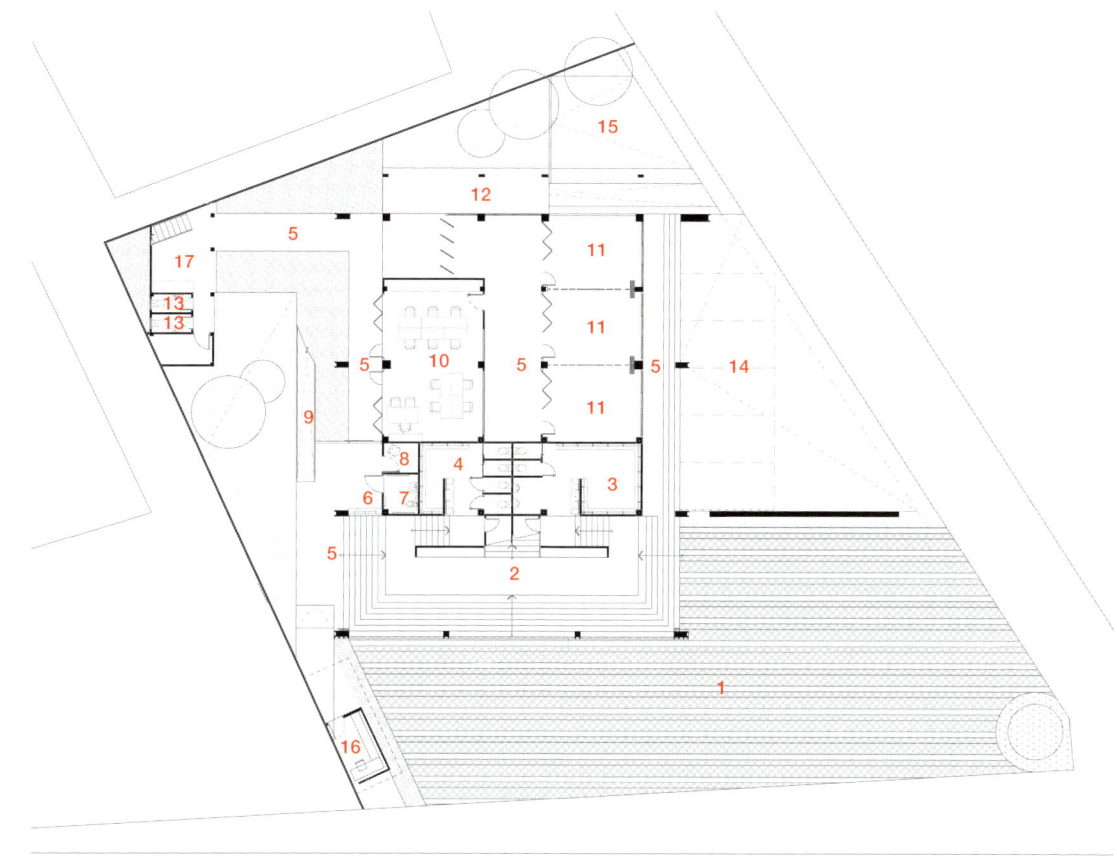

↑ Main praying area

1 COURTYARD
2 ENTRANCE HALL
3 MEN ABLUTION AREA
4 WOMEN ABLUTION AREA
5 CORRIDOR
6 DIFABLE ABLUTION AREA
7 DIFABLE TOILET
8 NURSING ROOM
9 RAMP
10 ADMINISTRATION OFFCE
11 CLASSROOM
12 TERRACE
13 TOILET
14 CAR PARKING
15 MOTORCYCLE PARKING
16 SECURITY POST
17 PANTRY

SITE PLAN

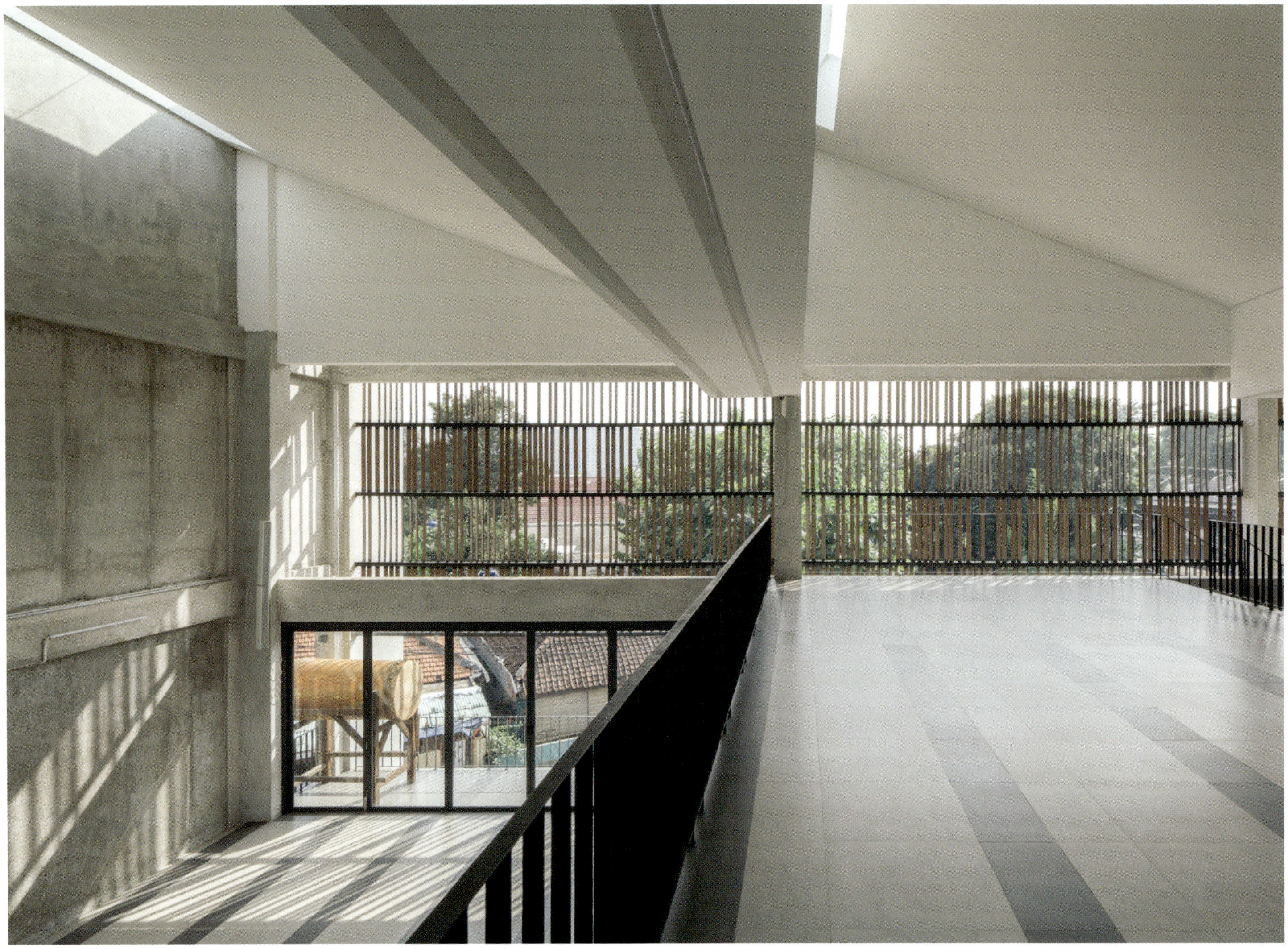

↑ Extension praying area

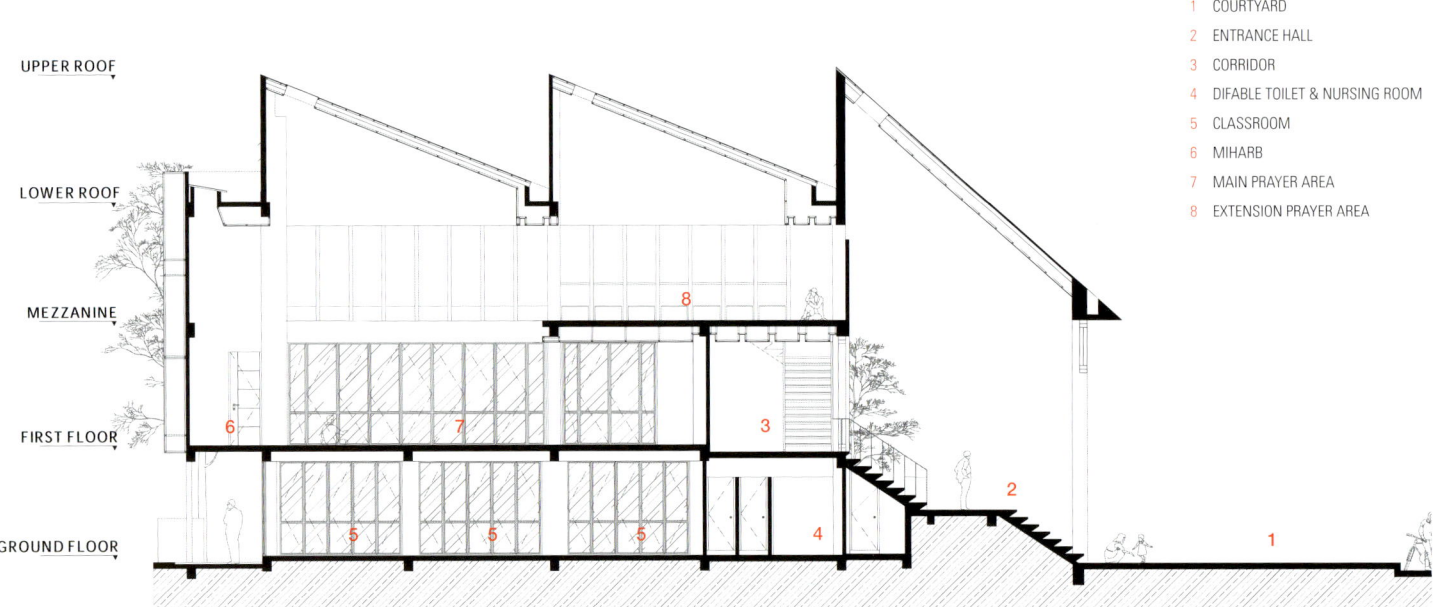

UPPER ROOF
LOWER ROOF
MEZZANINE
FIRST FLOOR
GROUND FLOOR
1 COURTYARD
2 ENTRANCE HALL
3 CORRIDOR
4 DIFABLE TOILET & NURSING ROOM
5 CLASSROOM
6 MIHARB
7 MAIN PRAYER AREA
8 EXTENSION PRAYER AREA

SECTION

← Skylight → Skylight

1 ENTRANCE HALL
2 MEN ABLUTION AREA
3 WOMEN ABLUTION AREA
4 CORRIDOR
5 DIFABLE ABLUTION AREA
6 DIFABLE TOILET
7 NURSING ROOM
8 RAMP
9 ADMINISTRATION OFFCE
10 CLASSROOM
11 TERRACE
12 TOILET
13 STORAGE
14 PANTRY
15 MAIN PRAYER AREA
16 MIHRAB
17 AUDIO ROOM
18 IMAM ROOM
19 EXTENSION PAYER AREA

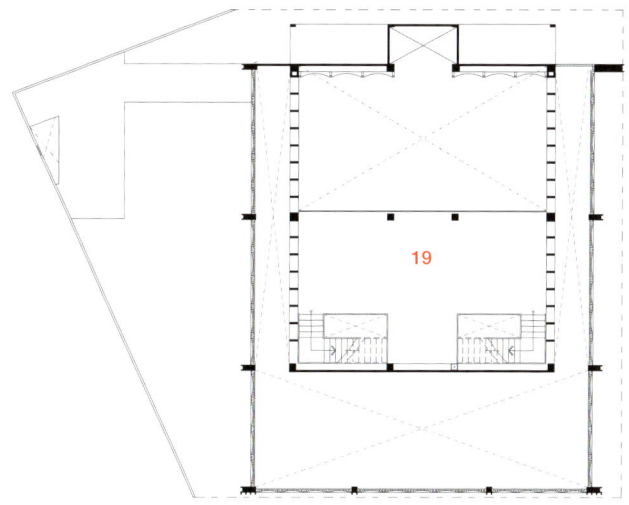

MEZZANINE FLOOR PLAN

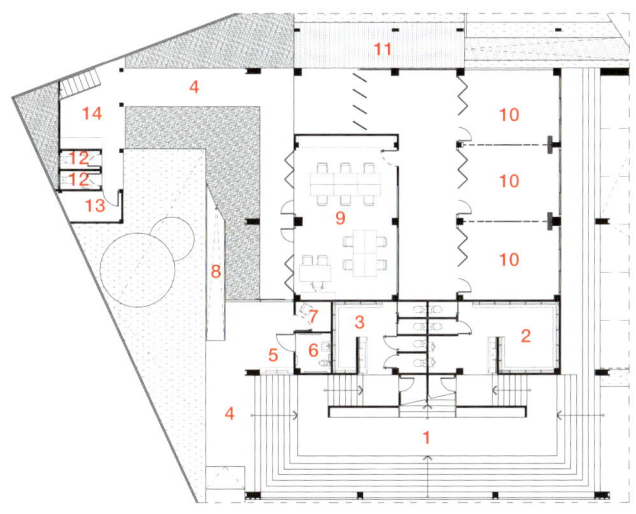

GROUND FLOOR PLAN

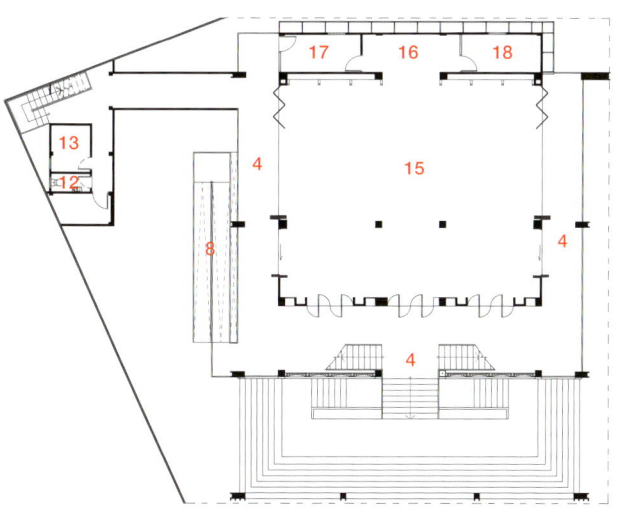

1ST FLOOR PLAN

109

발타사르 빌딩
BALTASAR BUILDING

ARCHITECT : SANTA-CRUZ ARQUITECTURA / JUAN ANTONIO SANTA-CRUZ GARCIA

BALTASAR BUILDING IS A COLLECTIVE HOUSING PROJECT that addresses the balance between heritage conservation and urban densification from the framework of the three pillars of sustainability.

SOCIAL SUSTAINABILITY

Strategy : A project transforming a sober 19th century neoclassical building in a state of ruin into affordable rental apartments. Two new floors providing private homes are added; bringing a new perspective to the building and the area that merges past and contemporary. A mixed residential model aiming to energize and diversify the social profile in a highly densified neighborhood.

Identity : Connecting with the city's motto "Priscas novissima exaltat et amor" meaning "to embrace and love the old and new". A strategy to preserve and put value on the emblematic elements of the existing building, adapting them to contemporary needs and domesticity.

Equity : Balancing the quality of rental and private housing, preventing the hierarchy of the property standards by creating an inclusive design that offers a response to different residents and needs.

Local crafts and industrial elements : Exploring ornamentation and traditional crafts through the new industrial technique of local manufacturers (such as steel and rope latticework) and a contemporary reinterpretation, in a constant dialogue with the artisan elements of the pre-existing building. In the building extension the rope lattice and the stainless-steel die-cut sheet lattice are performed by local artisans.

ENVIRONMENTAL SUSTAINABILITY

Biophilia, In order to recover the ecological relationship with nature and its benefits for well-being, new infrastructures related to the support of plants are added, as well as the use of natural colors and materials. On the top floor terraces, a garden of native species is placed with the aim of promoting biodiversity and reducing the heat island effect in the city.

Bioclimatism : There are several bioclimatic solutions to reduce energy consumption. Among them are the EIFS-Thermal Inertia combination, the renovation of windows with thermal break or the use of solar protection elements, such as traditional external blinds in the existing building and the use of latticework in the extended floors.

ECONOMIC SUSTAINABILITY

Achieved from energy efficiency bringing down energy consumption, resource efficiency, rehabilitating and taking advantage of an existing building, or local economy, reducing transport costs and contributing to improve the economy activity in the area.

Location Plaza González Conde, 2, Barrio del Carmen, Murcia, Spain **Use** Housing **Built area** 1.345,99m² **Gross floor area** 191,58m² **Completion** 2022 **Project manager** Juan Antonio Santa-Cruz Alemán, Juan Antonio Santa-Cruz García, Carmen Santa-Cruz García, Javier Esquiva López, Beatriz Lorente Martínez **Contractor** Guirao Sánchez y Sociedad, C.B **Photographer** David Frutos

← Building view

← Exterior & street view → Exterior & street view ↙ Exterior night view

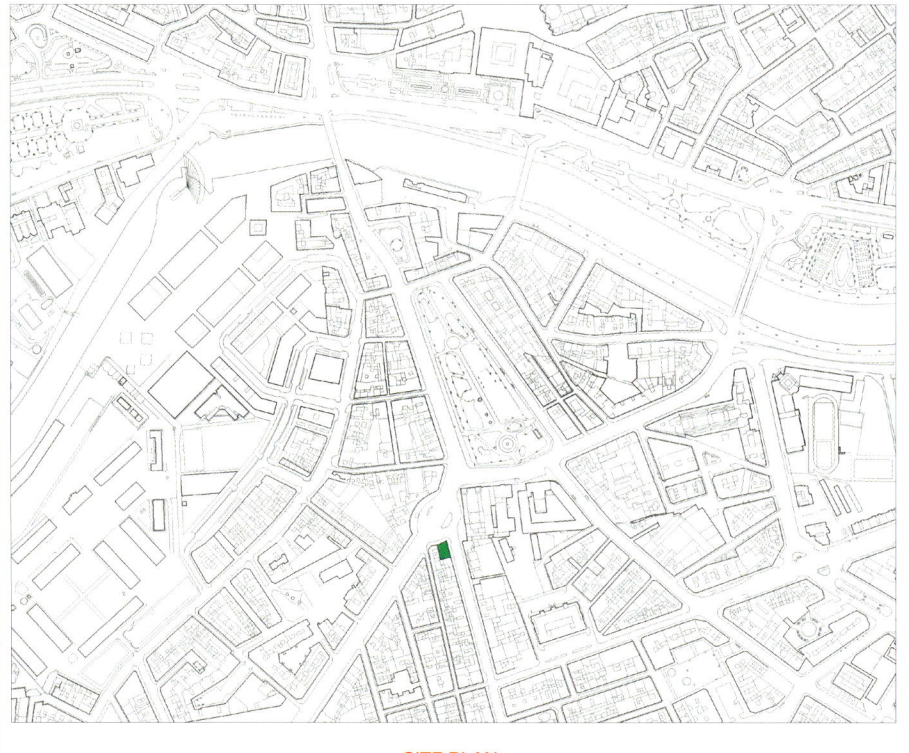

SITE PLAN

← Before construction → After construction

발타사르 빌딩은 세 축의 지속가능성을 반영해 전통과 보존, 도시 밀집화의 균형을 추구하는 공동 주택 프로젝트이다.

사회적 지속가능성
전략 : 낡은 19세기 신고전주의 건물을 저렴한 임대 아파트로 바꾸는 프로젝트이다. 2개의 층으로 이루어진 개인 주택이 추가되면서 과거와 현대가 어우러진 건물이 완성되었다. 혼합 주거 모델은 밀집화된 지역에 다양성의 요소와 활기를 더하는 것을 목표로 했다.
정체성 : "오래된 것과 새로운 것을 포용하고 사랑하자"는 의미의 "Priscas novissima exaltat et amor"라는 도시의 모토와 연결된다. 기존 건물의 상징적인 요소를 보존하고 가치를 부여하는 동시에 현대 사회의 주택 필요성을 충족시키는 전략이다.
형평성 : 임대 및 개인 주택 품질의 수준을 균형 있게 맞추고 다양한 거주자의 요구를 충족시키는 포괄적인 디자인으로 부에 따른 위계를 약화시키고자 했다.
지역 공예 및 산업 요소 : 기존의 장인적 요소를 유지하면서 지역 제조업체의 새로운 산업 기술 (강철 및 로프 격자 작업) 및 현대적 재해석을 통해 새로운 장식 및 전통 공예를 시도했다. 로프 격자와 다이 커팅한 스테인리스 강판을 이용한 건물 확장 작업은 현지 장인이 담당했다.

환경적 지속가능성
바이오필리아, 자연과 생태적 관계를 회복하고 웰빙 라이프의 이점을 누리기 위해 기본 인프라에 식물 요소를 추가할 뿐 아니라 자연스러운 색상과 재료를 사용하였다. 생물 다양성을 촉진하고 도시의 열섬 효과를 줄이기 위해 자생 식물로 구성된 정원이 최상층 테라스에 만들어 졌다.
생태기후적 요소 : 에너지 소비를 줄이기 위해 생태기후적 요소가 적용됐다. 이는 외단열미장 공법(EIFS), 단열 창, 기존 건물의 일반 외부 블라인드 역할을 하는 자외선 차단 요소, 확장 공간의 격자 바닥을 포함했다.

경제적 지속가능성
에너지 소비 감소를 통한 에너지 효율성, 자원 효율성, 기존 건물이나 지역 경제의 재건 및 이점 활용을 통해 운송 비용 절감 및 해당 지역의 경제 활동 개선에 공헌했다.

← Two new floors providing private homes

↑ Steel and rope latticework view   ↳ Two new floors night view

ELEVATION

↑ 1st floor entrance & lobby ↙ Stair ↓ Stair ↘ Wall design

↑ Living room   ↙ Living room   ↳ Kitchen

↱ 4th floor livingroom
← Staircase
↳ 4th floor livingroom

← Top floor veranda　→ Top floor veranda

1 RIBBED SANDWICH PANEL
2 POINT ANCHORS
3 ANTI-ROOT GEOTEXTILE
4 FINISH : PLANTER
5 ANTI-ROOT WATERPROOFING SHEET
6 STYRODUR
7 GEOTEXTILE
8 DRAINAGE LAYER WITH HOLES FACING UP WITH GEOTEXTILE
9 FINISH : MICROCEMENT
10 REINFORCED SELF-LEVELING
11 FINISH : MICROCEMENT
12 ARLITA FILLING
13 CONTINUOUS SELF-LEVELING 7cm
14 UPN 60 PUNCTUAL GUIDE SLIDE ANCHOR
15 ROCK WOOL
16 PLASTERBOARD+PKB2
17 PUNCTUAL GUIDE SLIDE ANCHOR PLATE
18 AQUAPANEL
19 SELF-PROTECTED ASPHALT SHEET
20 FINISHES : MARBLE, PORCELAIN, TILE
21 FINISH : WOODEN FLOORING
22 BATTEN SYSTEM 3cm
23 POINT ANCHORS
24 5mm ELASTIC MORTAR ARMED WITH MESH
25 FILLED WITH SIKAFLEX
26 POLYDROS THERMAL INSULATION
27 THEY IMPACT
28 CONTINUOUS SELF-LEVELING 7cm
29 WEBER THERM INSULATING MORTAR FINISH + WEBER THERM CLIMA 7cm

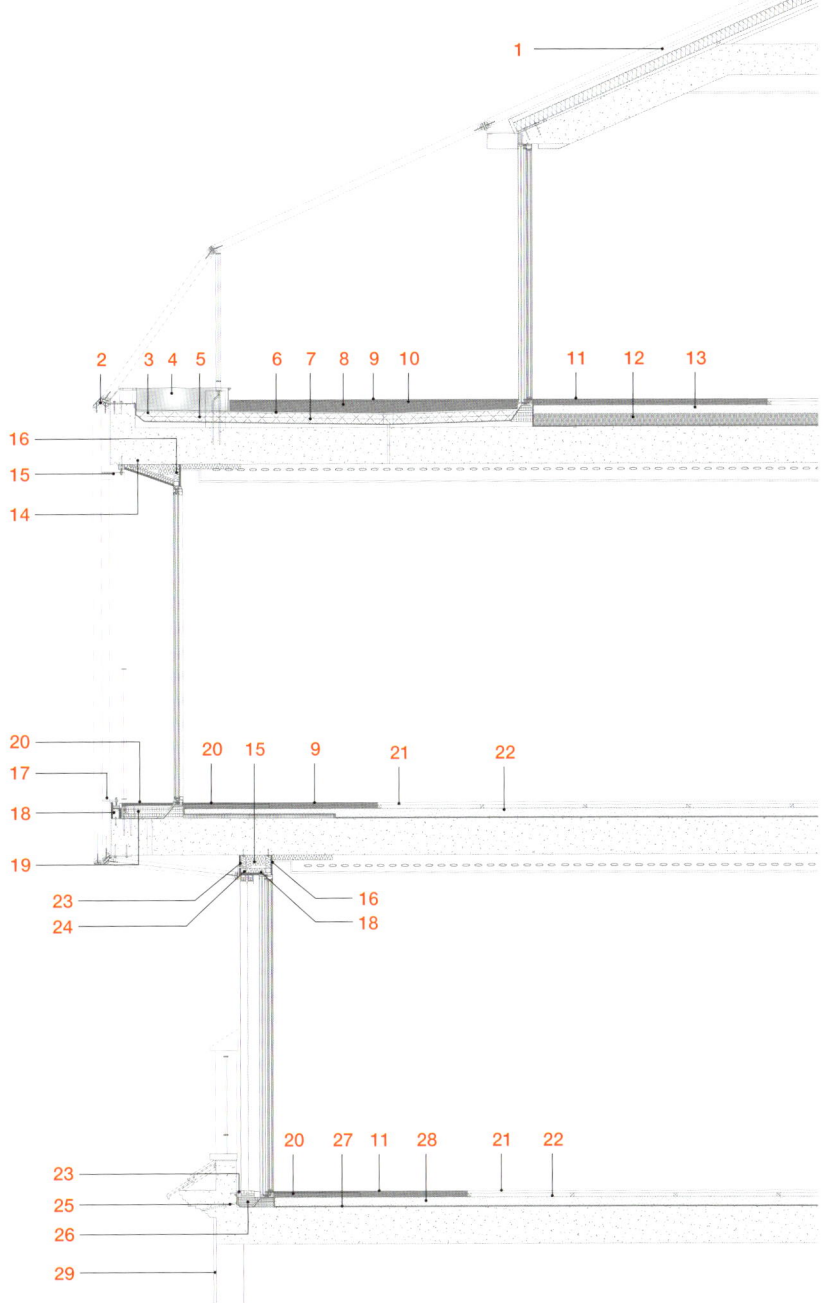

SECTION DETAIL

← Bathroom   ↙ Bathroom   → 5th floor stair

1 ELEVATOR   3 LIVING ROOM   5 BEDROOM   7 DINING ROOM   9 VERANDA
2 ELEVATOR HALL   4 KITCHEN   6 TOILET   8 BALCONY   10 STUDY ROOM

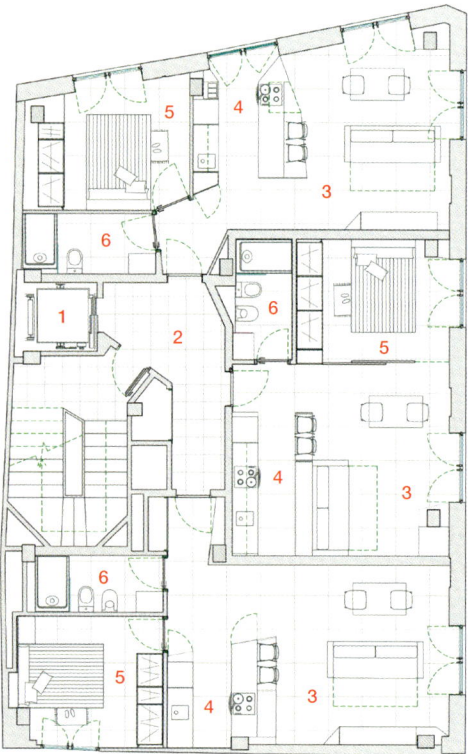

TYPICAL FLOOR PLAN

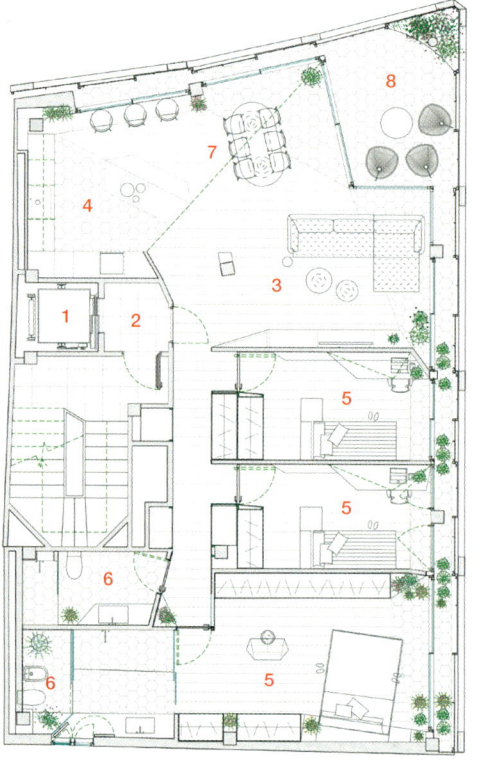

4TH FLOOR PLAN

↴ Top floor veranda   ↵ Top floor veranda   → 5th floor balcony

5TH FLOOR PLAN

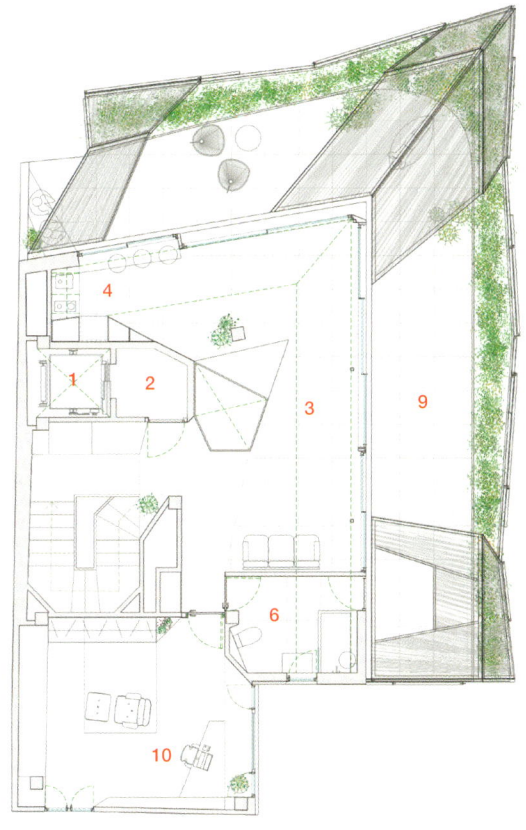

6TH FLOOR PLAN

121

그레이하운드 하우스
GREYHOUND HOUSE

ARCHITECT : MURADO & ELVIRA ARQUITECTOS / CLARA MURADO, JUAN ELVIRA

THE HOUSE IS BUILT IN A LONG AND NARROW PLOT in a quiet residential area in Madrid, unused for years due to its unusual proportions. It accommodates the floor plan requirements inside a volume only 4 meters wide, resulting in a thin and compact prism.

On each level, small outdoor spaces are inserted, allowing for a rich spatial experience while keeping the necessary privacy in a dense residential neighborhood. This sequence encompasses a courtyard on the basement level, a garden with a swimming pool on the ground floor and a double-height terrace on the first floor expand and complement the tight interior of the house. Thus, the usual sequence of stacked levels is challenged and turned into a dynamic vertical landscape.

The basement is illuminated through a dug patio that occupies the entire front of the plot. Over this void, a bridge connects the street directly to the garden and the entrance of the house.

The ground floor opens into the garden and the pool, which runs longitudinally along the edge. On this level, the living room and kitchen have large windows that allow the continuity of these spaces. Large perforated metal sliding shutters nuance the light. The circular cutouts of the sliding doors become the identity of the house.

On the first level, the main bedroom area occupies the whole floor, consisting of a bedroom, a walk-in closet and a large bathroom. The bathroom connects to the main terrace on the front of the volume. This terrace is enclosed by double-height walls with some openings to keep privacy and enjoy views, like a room without a roof. A spiral staircase floats over this space, connecting the second floor to the solarium on the roof, where you can enjoy the wide views of the entire area.

Location Madrid, Spain **Use** Residential **Site area** 266m² **Built area** 280m² **Design team** Christine Gutiérrez Chevalier, Victoria Bosch **Contractor** APRO **Photographer** Imagen Subliminal

이 집은 마드리드의 조용한 주거 지역에 길고 좁은 부지에 세워졌으며, 익숙하지 않은 조형적 비율로 인해 수년간 방치되어 있었다. 평면도를 따라 내부 폭이 4m로 좁고 길쭉한 형태이다.

각 층에는 야외 공간이 조성되어 공간을 다층적으로 만들어 주며, 밀집된 주거 지역에서 사생활을 보호해 준다. 지하에는 안뜰, 1층에는 수영장과 정원, 이중 높이의 테라스를 구성해 공간을 확장시키고 좁은 내부를 보완해 준다. 즉, 층간 구성을 비틀어 역동적인 수직적 구조를 연출한다.

부지 전체가 깊이 파인 듯한 테라스로 이루어진 지하층이 시선을 사로잡고 넓은 공간 위로 다리가 놓여 복도와 정원, 집의 입구로 곧장 연결된다.

1층은 가장자리를 따라 세로로 이어지는 정원과 수영장으로 연결된다. 또한 거실과 주방에는 커다란 창이 있어 공간의 연속성을 확보했다. 천공을 가진 대형 금속 슬라이딩 셔터를 통해 빛이 은은하게 들어오게 했다. 미닫이 문에 장식한 원형 컷아웃 장식은 집의 정체성을 드러낸다.

1층 전체를 차지하는 안방 침실에는 침대와 대형 옷장, 대형 욕실이 구성되어 있다. 욕실은 주택 전면의 메인 테라스로 연결된다. 이중 높이의 벽으로 둘러싸인 테라스에는 트인 부분이 있어 사생활이 보호되면서도 지붕이 없는 방처럼 전망을 즐길 수 있다. 이 공간 위로 조성된 나선형 계단은 2층을 옥상의 일광욕실과 연결해 주변 지역의 탁 트인 전망을 감상 할 수 있다.

← Exterior & main entrance view   → Exterior night view

↱ View from above
↳ Building view

↑ Front view   ↳ Patio & bridge

AXONOMETRIC

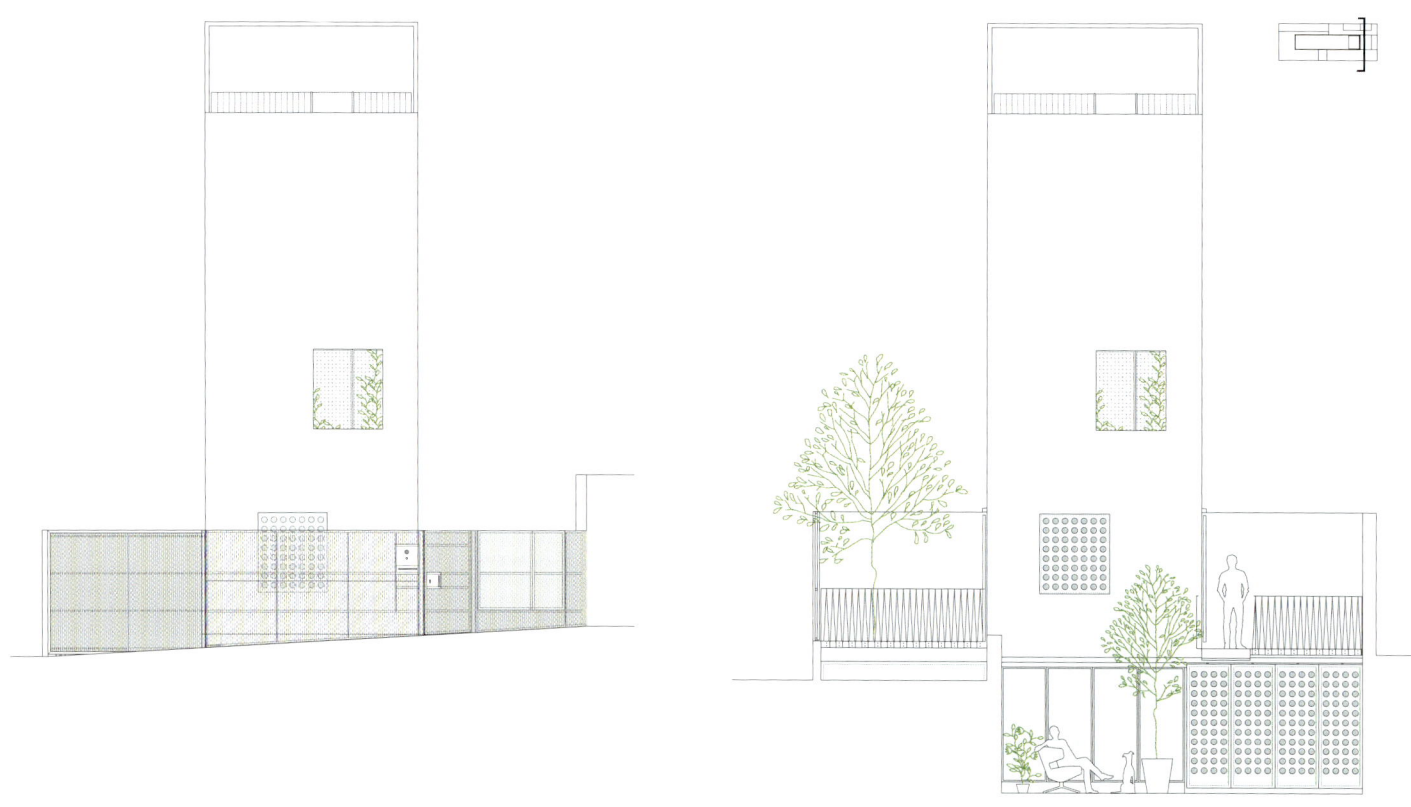

↑ Swimming pool on the ground floor

EAST ELEVATION

← Garden with a swimming pool on the ground floor   → The circular cutouts of the sliding doors

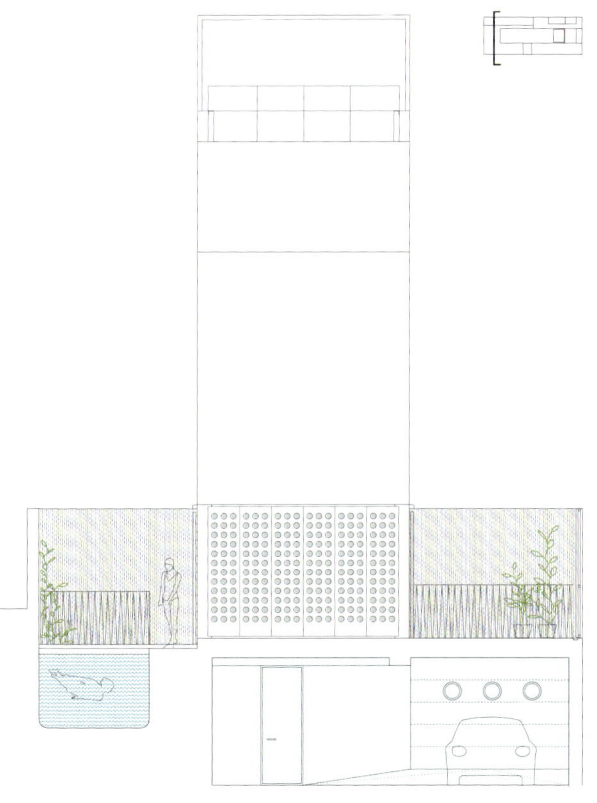

WEST ELEVATION

↑ Basement level parking lot door

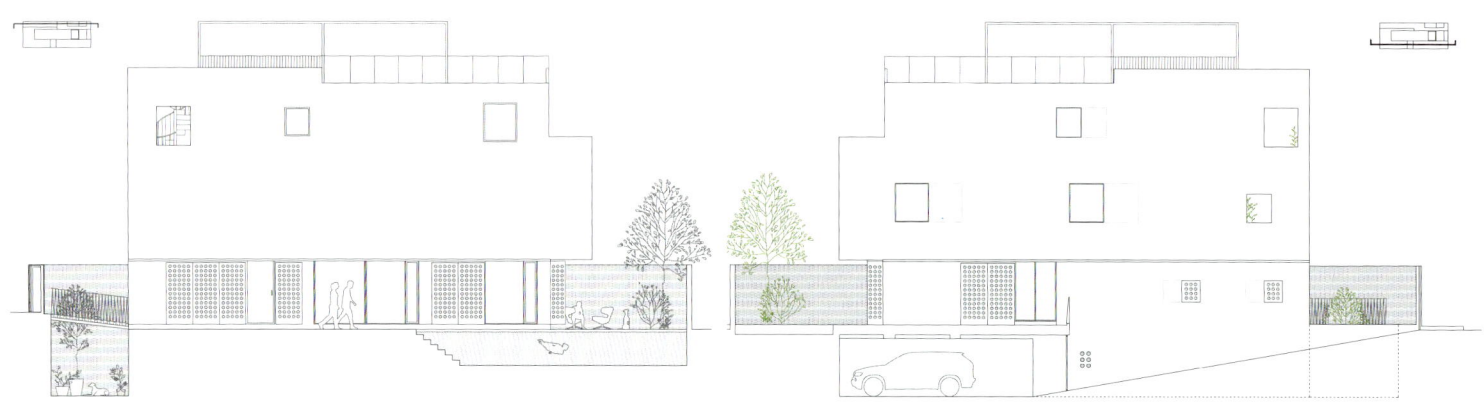

NORTH ELEVATION

SOUTH ELEVATION

← Spiral staircase → View up in the garden

LONGITUDINAL SECTION

129

↑ Ground floor stair case ← Ground floor stair case

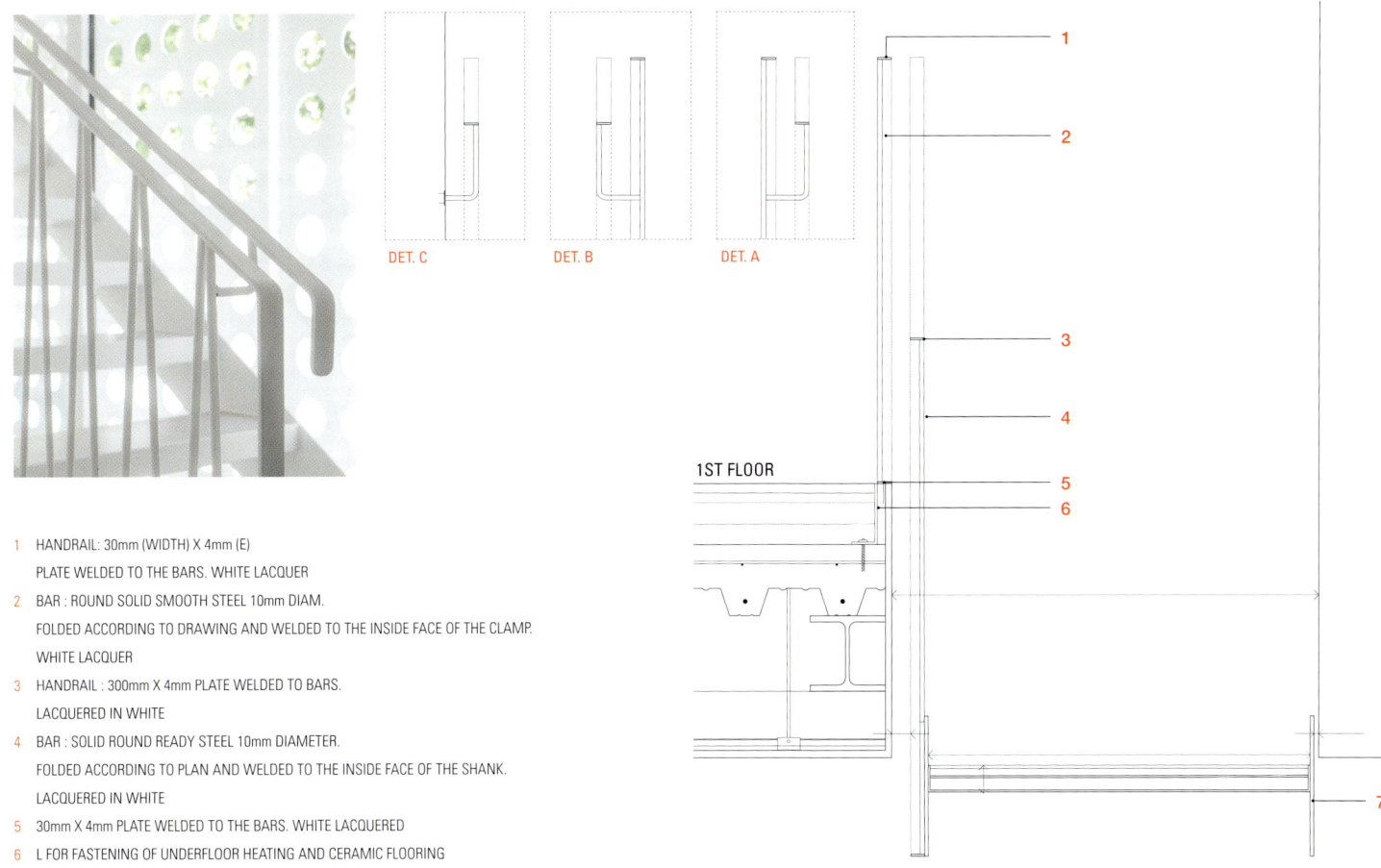

1. HANDRAIL: 30mm (WIDTH) X 4mm (E)
   PLATE WELDED TO THE BARS. WHITE LACQUER
2. BAR : ROUND SOLID SMOOTH STEEL 10mm DIAM.
   FOLDED ACCORDING TO DRAWING AND WELDED TO THE INSIDE FACE OF THE CLAMP.
   WHITE LACQUER
3. HANDRAIL : 300mm X 4mm PLATE WELDED TO BARS.
   LACQUERED IN WHITE
4. BAR : SOLID ROUND READY STEEL 10mm DIAMETER.
   FOLDED ACCORDING TO PLAN AND WELDED TO THE INSIDE FACE OF THE SHANK.
   LACQUERED IN WHITE
5. 30mm X 4mm PLATE WELDED TO THE BARS. WHITE LACQUERED
6. L FOR FASTENING OF UNDERFLOOR HEATING AND CERAMIC FLOORING
7. STAIR STRINGER 8mm STEEL SHEET, WHITE LACQUERED

HANDRAIL DETAIL

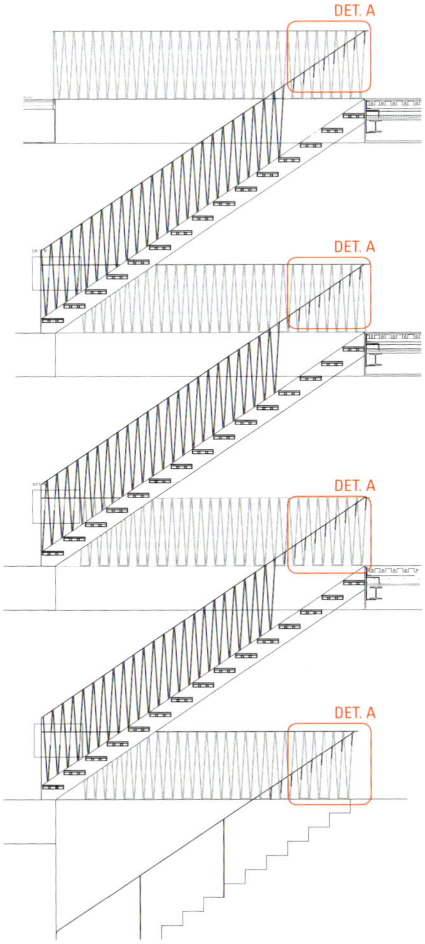

STAIRCASE SECTION

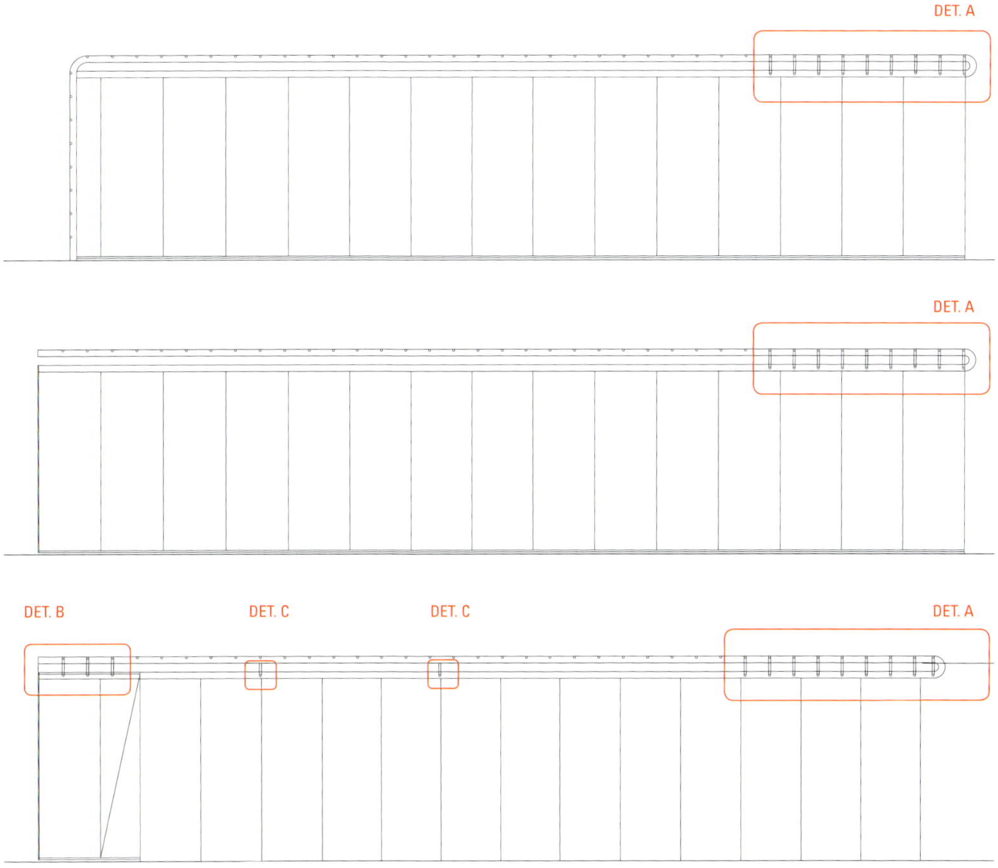

STAIRCASE PLAN

↑ Ground floor living room   ↙ 1st floor hallway   ↘ Terrace on the main bathroom

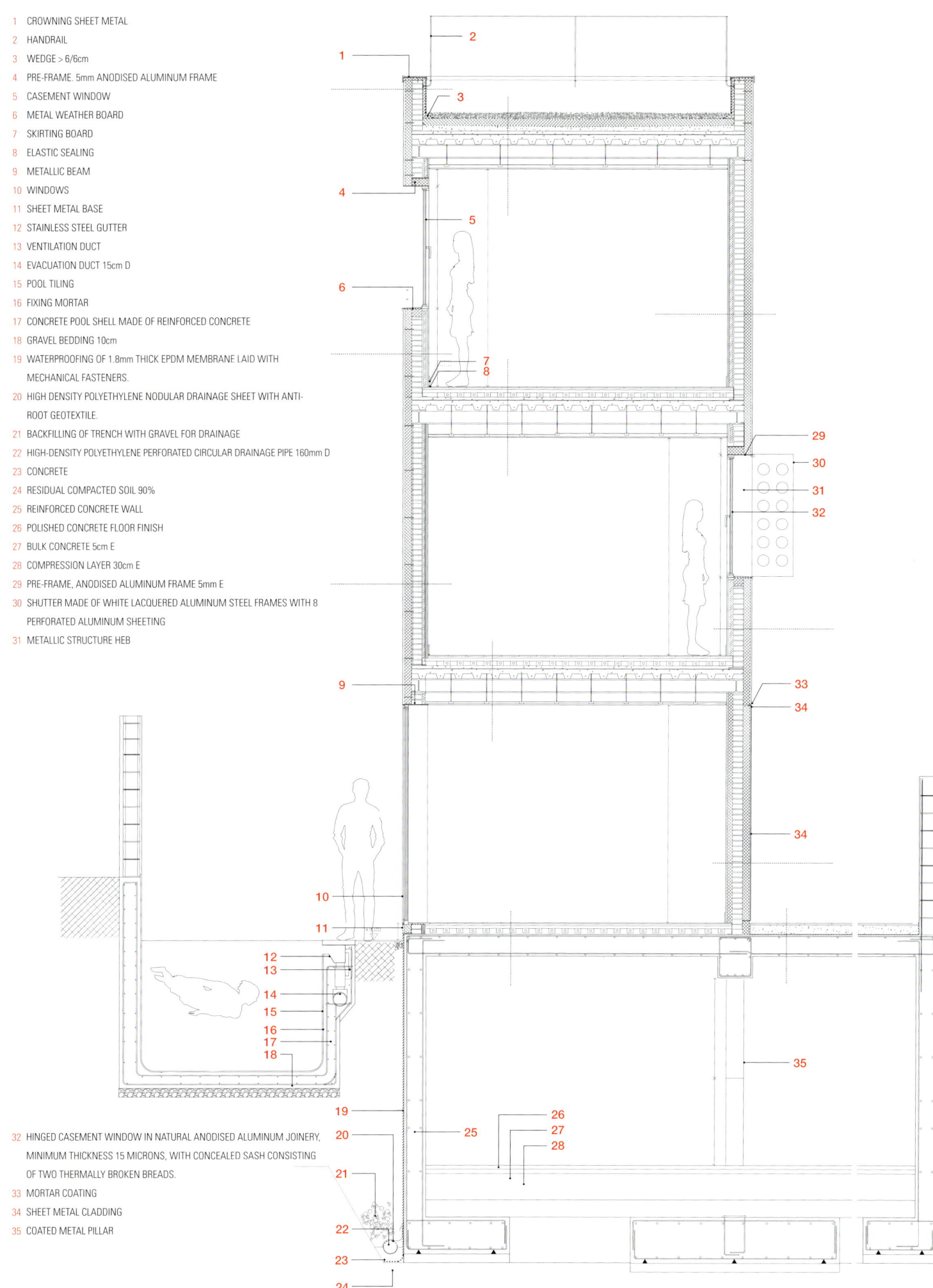

1 CROWNING SHEET METAL
2 HANDRAIL
3 WEDGE > 6/6cm
4 PRE-FRAME. 5mm ANODISED ALUMINUM FRAME
5 CASEMENT WINDOW
6 METAL WEATHER BOARD
7 SKIRTING BOARD
8 ELASTIC SEALING
9 METALLIC BEAM
10 WINDOWS
11 SHEET METAL BASE
12 STAINLESS STEEL GUTTER
13 VENTILATION DUCT
14 EVACUATION DUCT 15cm D
15 POOL TILING
16 FIXING MORTAR
17 CONCRETE POOL SHELL MADE OF REINFORCED CONCRETE
18 GRAVEL BEDDING 10cm
19 WATERPROOFING OF 1,8mm THICK EPDM MEMBRANE LAID WITH MECHANICAL FASTENERS.
20 HIGH DENSITY POLYETHYLENE NODULAR DRAINAGE SHEET WITH ANTI-ROOT GEOTEXTILE.
21 BACKFILLING OF TRENCH WITH GRAVEL FOR DRAINAGE
22 HIGH-DENSITY POLYETHYLENE PERFORATED CIRCULAR DRAINAGE PIPE 160mm D
23 CONCRETE
24 RESIDUAL COMPACTED SOIL 90%
25 REINFORCED CONCRETE WALL
26 POLISHED CONCRETE FLOOR FINISH
27 BULK CONCRETE 5cm E
28 COMPRESSION LAYER 30cm E
29 PRE-FRAME, ANODISED ALUMINUM FRAME 5mm E
30 SHUTTER MADE OF WHITE LACQUERED ALUMINUM STEEL FRAMES WITH 8 PERFORATED ALUMINUM SHEETING
31 METALLIC STRUCTURE HEB

32 HINGED CASEMENT WINDOW IN NATURAL ANODISED ALUMINUM JOINERY, MINIMUM THICKNESS 15 MICRONS, WITH CONCEALED SASH CONSISTING OF TWO THERMALLY BROKEN BREADS.
33 MORTAR COATING
34 SHEET METAL CLADDING
35 COATED METAL PILLAR

DETAIL SECTION

← Staircase, 2nd floor   → Staircase

1 ENTRANCE
2 COURTYARD
3 SWIMMING POOL
4 KITCHEN
5 TOILET
6 DINING ROOM
7 LIVING ROOM
8 SLIDING DOORS
9 GARDEN
10 PARKING LOT
11 TERRACE
12 BATHROOM
13 STAIR
14 STUDY SPACE
15 BEDROOM
16 SOLARIUM ON THE ROOF

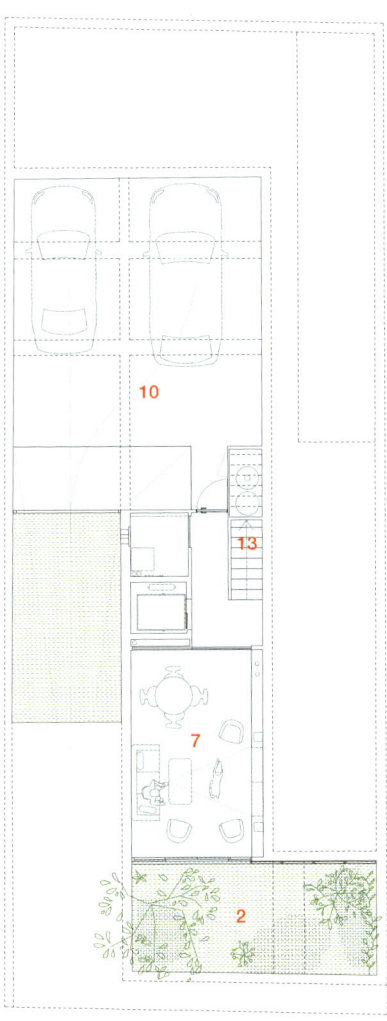 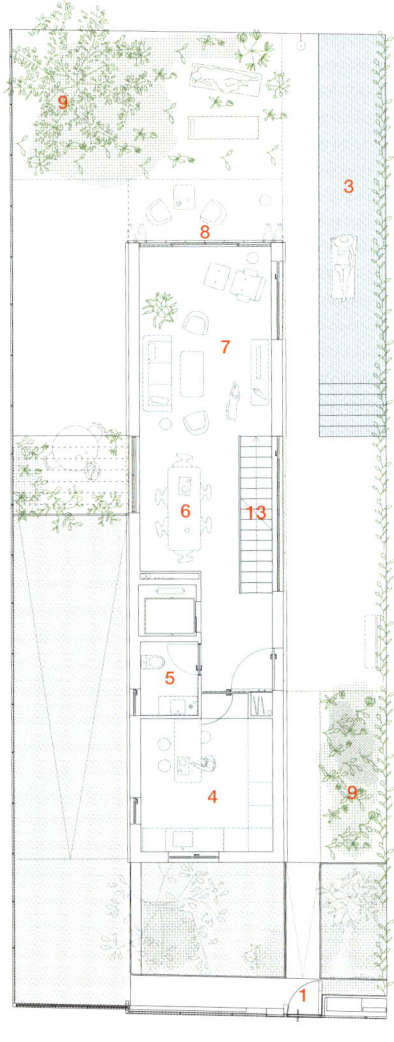

BASEMENT FLOOR PLAN   **GROUND FLOOR PLAN**

← Staircase → Staircase & study space

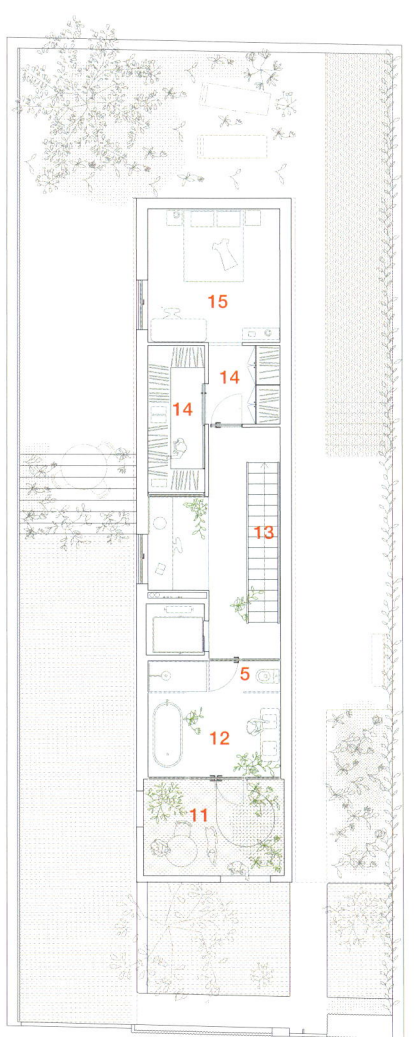
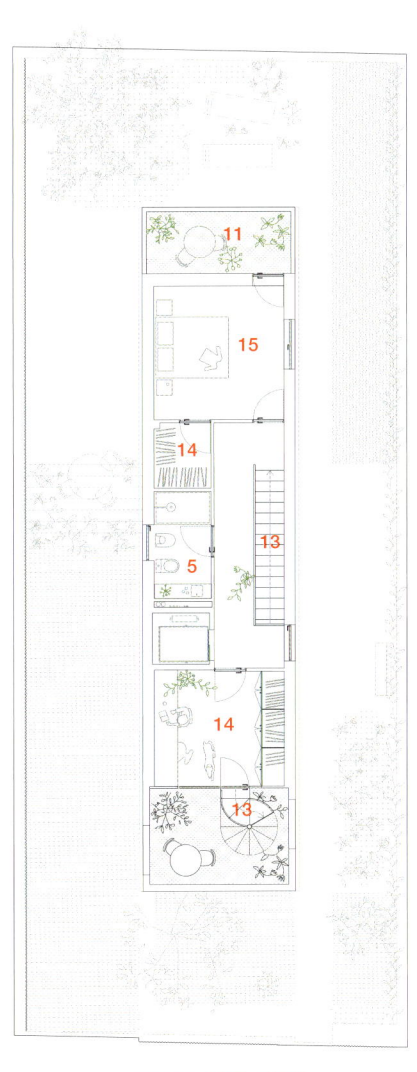
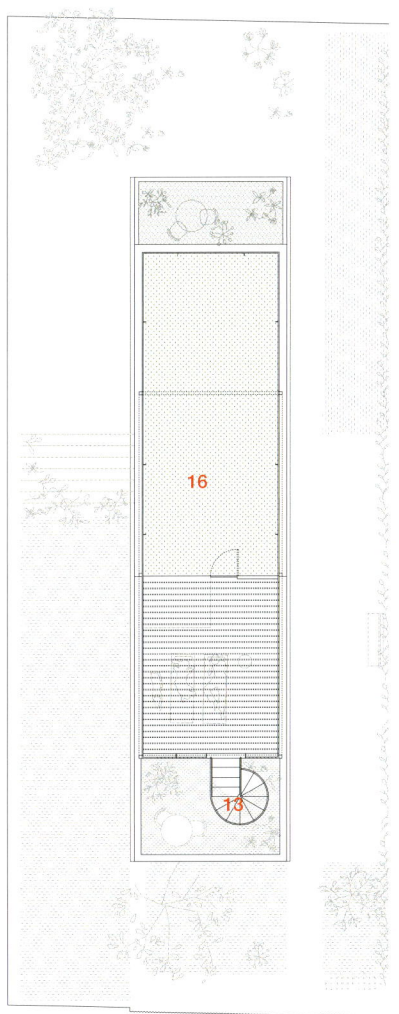

1ST FLOOR PLAN **2ND FLOOR PLAN** **ROOF TOP FLOOR PLAN**

쿠보미 아파트먼트

KUBOMI APARTMENT

ARCHITECT : ORGANIC DESIGN INC. / HIDEO KUMAKI

ORGANIC DESIGN INC. HAD TO PLAN TO VALUE THE ATTRACTIVENESS that conventional apartments do not have by client requests , In design, they would be encouraging residents to be considerate neighbors, as single people living under the same roof a friendly place, with communication preventing isolation. This inspired the idea of a central courtyard, which is not necessarily unusual but still provides a place for residents to linger for a chat. Kubomi means "pocket," among other things, so it's a fitting description for this gathering place. Besides fostering these ties, the courtyard also links the apartments to the neighborhood, forming a space that can replace the isolation of solitary residents with a sense of being part of a community.

The project start date in August 2020 during the pandemic meant that they paid closer attention to certain topical issues: ideal lives/lifestyles for single people, how people share things with each other, and how to support this sharing. The decision to build studio apartments responded to the neighborhood's high concentration of medical professionals and the fact that the pandemic was spreading. A two-story structure here would suit the lot shape and size, their preference for wooden construction, and the need to meet building codes. Quiet units are an advantage when tenants include medical professionals and workers with irregular hours, so this was a priority in planning and construction of the floors, walls, and ceilings, while forgoing balconies, electric clothes dryers, and modular bathrooms (in favor of modular showers with benches) ensured maximum space inside and room for other fixtures, such as hygienic washbasins.

Setting the scene not only for tenant interaction but also community communication is the carefully designed shape of the inviting, accessible courtyard. Formed by dividing the building into two separate structures with curved walls, it creates an inlet from the street that tempts people to approach and slip in. Another touch that draws people in is the seamless merging of the road with the asphalt of the courtyard.

In consideration of the pandemic, ease of expanding shared spaces was not addressed in the design of the apartments, but we did seek to avoid people feeling isolated—another significant social issue for some time by planning a sense of privacy that also allows tenants to be mindful of each other. They hope that in some way, the apartments help break down excessive uniformity in society and indifference to one's neighbors.

Location Saitama, Japan **Use** Apartment **Site area** 329.73m² **Building area** 199.84m² **Gross floor area** 356.72m² **Completion** 2023 **Project manager** Hideo Kumaki, Rei Maki **Contractor** Ishizue Column Co.,Ltd. **Photographer** Yukinori Okamura

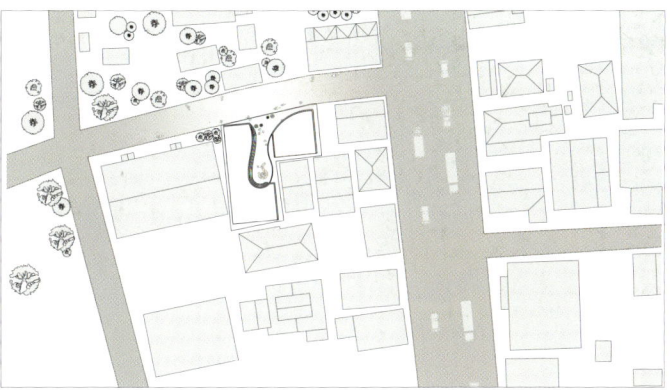

SITE PLAN

← Patio view   → Night view

↑ Exterior view

오가닉 디자인은 기존 아파트에 없는 매력적인 공간을 원하는 고객의 요청을 만족시키고자 했다. 설계를 할 때 한 지붕 아래에서 홀로 살아가는 이웃과의 거리감을 좁히는 친밀한 공간을 설계해 누구도 소외되지 않는, 소통하는 공간을 만들고자 하였다. 그래서 완전히 특이한 아이디어는 아니지만, 거주민들이 모여서 담소를 나눌 수 있는 안뜰을 중앙에 조성하였다. 이는 무엇보다 '주머니' 라는 뜻의 쿠보미라는 이름에서 이 모임 장소가 갖는 중요성을 알 수 있다. 안뜰은 유대를 강화하는 공간인 동시에 아파트와 거주민을 연결해 고독한 거주자가 고립된 공간을 하나의 공동체로 묶는 역할을 해준다.

팬데믹 기간인 2020년 8월에 시작된 이 프로젝트는 독신자를 위한 이상적인 일상과 라이프스타일, 거주민 간의 물건 공유 및 공유를 지원하는 방법들에 대해 특히 관심을 두었다. 원룸식 구조로 결정한 것은 팬데믹이 점차 확산되는 가운데 전문 의료진이 많은 거주민의 필요성 때문이다. 또한 건축 규정을 고려해 부지 형태와 크기에 적합한 복층 구조의 목조 건물로 결정했다. 근무 시간이 불규칙한 전문 의료진 및 근로자와 같은 주민들을 위해 바닥, 벽 및 천장 설계를 할 때 조용한 공간을 조성하는 것을 우선순위에 두었고, 발코니, 전기 의류 드라이어, 모듈식 욕실(벤치가 있는 모듈형 샤워실) 등을 위해 내부 공간을 최대한 확보하고 위생적인 세면대와 같은 기타 설비 공간도 확보하였다.

개별 거주민뿐 아니라 전체 커뮤니티의 원활한 의사소통을 위해 매력적인 안뜰 공간을 쉽게 접근할 수 있도록 조성했다. 두 개의 개별 구조로 나누어진 건물의 외벽은 곡선을 이루며 사람들이 거리에서 자연스럽게 안으로 들어올 수 있는 통로를 연출한다. 또한 도로에서 안뜰로 매끄럽게 이어지도록 설계해 접근성을 향상시켰다.

이 아파트를 설계할 때 팬데믹을 고려해 공간을 용이하게 확장하는 것에는 크게 주안점을 두지 않았으며, 프라이버시를 지키면서도 일정 기간 사회적 문제가 되었던 사람들의 고립감을 완화시키기 위해 거주민들이 서로에게 관심을 가질 수 있는 공간으로 만들었다. 이 아파트를 통해 어떤 방식으로는 지나치게 획일화된 도시의 건물과 이웃에 대한 무관심이 줄어들기를 바란다.

↑ Patio

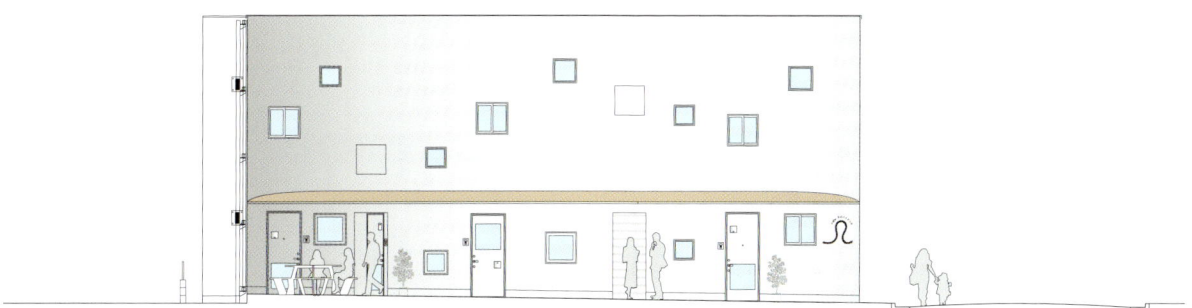

PATIO ELEVATION

NORTH ELEVATION

↑ Community table

SECTION

1. INTERIOR BOARDS ON EXTERIOR WALLS MUST BE RAISED UP TO THE WEATHERBOARD
2. EXTERIOR WALLS 2(APPROACHING) / FIRE CIRTIFICATION / COATING / T5 SPACER / T12 WONDER BOARD / 15X45 FURRING STRIPS, T20 VENT-LAYER / WATERPROOF-BREATHABLE-SHEET / T12.5 BENDING VENEERS / 105X105 COLLUMNS / 27X1 05@500 STUDS, T100 GLASS WOOL 24K
3. EAVES / ASPHALT SINGLE ROOF WOODSHAKE MARCHIKA / WATER PROOF VENEERS / 90X45 @455 RAFTERS / TOP COVER : ALUMINUM PLATE FOR DRAINER / VENTIATION CAPS / CEILING : WP COAT (ACRYLIC PLASTER) / LED LIGHT
4. ROOF : SHEET WATER PROOFING / FIREBRAND CERTIFICATION / VT COATING / T18 STRUCTURAL VENEERS / 30*40 RAFTERS VENTIATION GRILLS / OVERHEAD VENTILATION SYSTEM
6. STAIRCASE WALLS / FIRE CIRTIFICATION / COATING / T5 SPACER / T12 WONDER BOARD / 15X45 FURRING STRIPS, T20 VENTLAYER /WATER-PROOF-BREATHABLE-SHEET / T12.5 BENDING VENEERS / 105X105 COLLUMNS / 27X105 @500 STUDS, T100 GLASS WOOL 24K
7. VENTILATION CAP
8. EXTERIOR WALLS 1 / FIRE CIRTIFICATION / COATING / T5 SPACER / T12 WONDER BOARD / 15X45 FURRING STRIPS, T20 VENTLAYER / WATERPROOF-BREATHABLE-SHEET / T12.5 BENDING VENEERS / 105X105 COLLUMNS / 27X105 @500 STUDS, T100 GLASS WOOL 24K
9. FRP WATER PROOF STSTEM
10. LIGHTING WITH SENSER
11. COMMUNICATION BORAD
12. HOOKS FOR CLEANING TOOLS
13. TREAD : T60 WOOD BOARD + ANTISEPTIC PAINT / STRINGER : T9 STEEL + HOT DIP GALVANIZING

↑ Entrance hall & Staircase

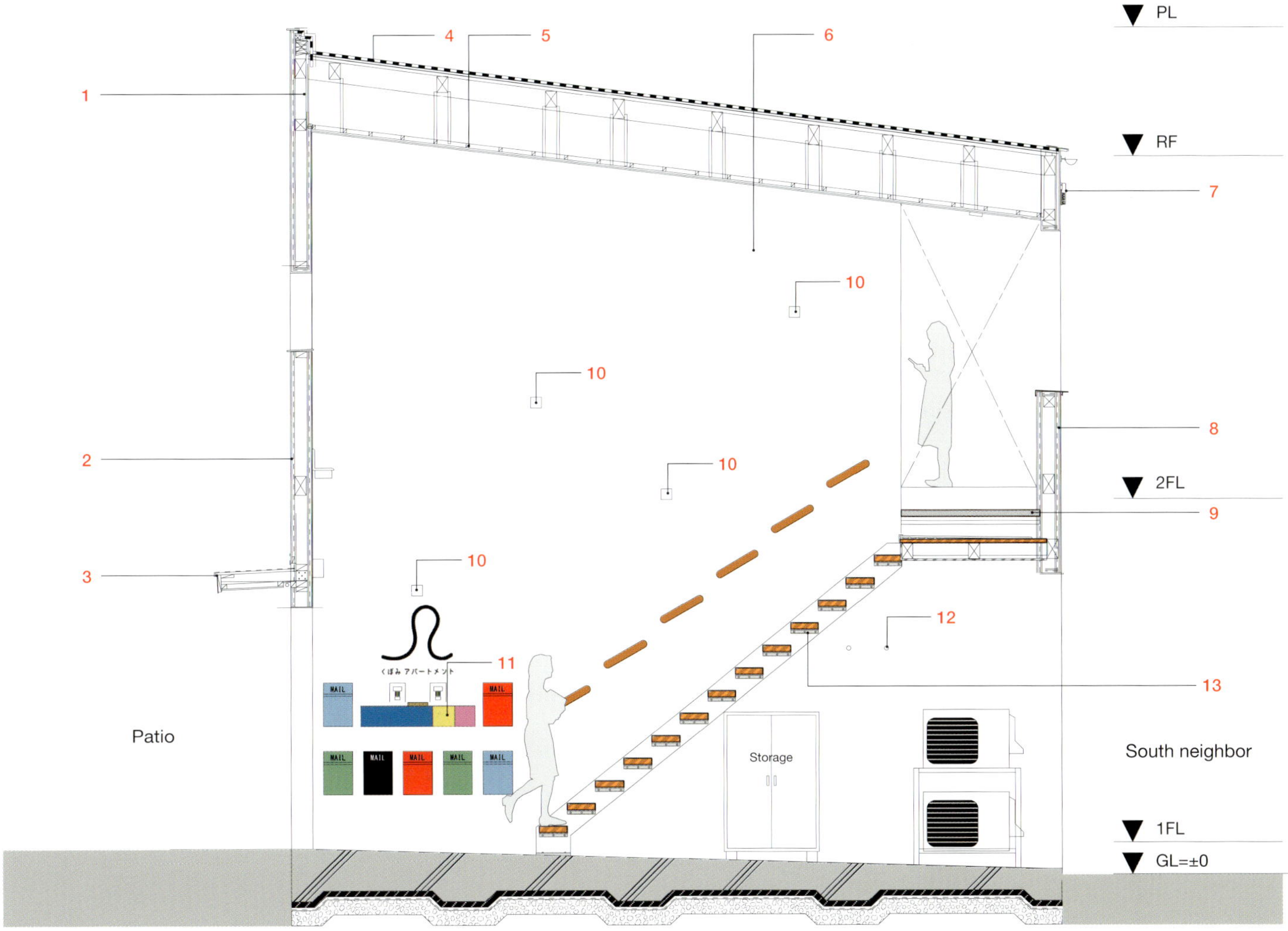

SECTION DETAILS OF STAIRCASE

↑ Residence unit   ↙ Colorful unit

1 KUBOMI PATIO
2 ENTRANCE
3 RESIDENCE UNIT
4 ENTRANCE HALL
5 STAIRCASE

SECTION

↑ Residence unit ⌐ Interior view ⌐ Interior view

1ST FLOOR PLAN

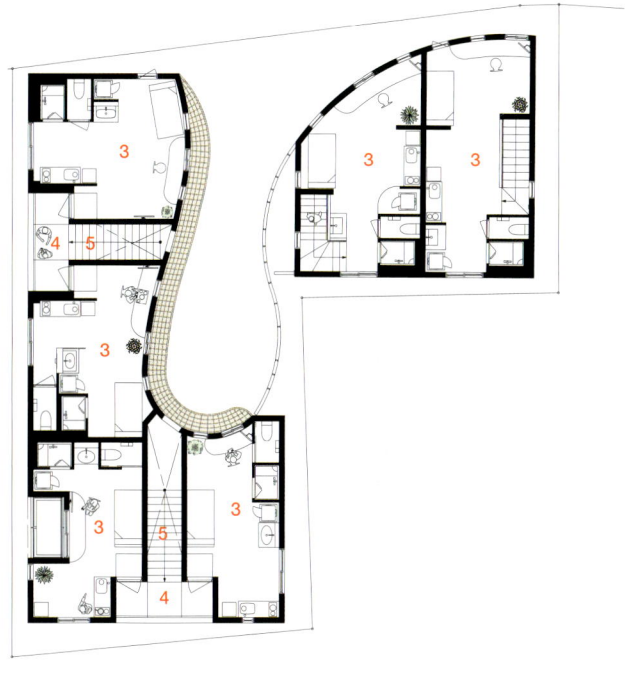

2ND FLOOR PLAN

에이치에스77

HS77

ARCHITECT : VON M / DENNIS MUELLER, MATTHIAS SIEGERT

LAND PRICES, CONSTRUCTION COSTS, and the need for a living/working balance, which became apparent once again in the Covid pandemic, lead to two houses—dug like identical twins into a steeply rising slope in the south of Stuttgart—whose floor plans allow for different forms of living today and in the future.

The semi-detached house takes up the eaves height and building depth of the neighboring house and formally plays out the theme of prefabrication in every detail. Because up to three stories had to be dug into the slope and in order to keep planning and construction costs low, a catalog of simple details and joints was developed using precast concrete elements. The disadvantages of the material (CO_2 emissions) are compensated by a long-term flexible use, which allows for different floor plan variants in the expansion and permits conversions. A cruciform floor plan scheme allows for different options of interior spatial joining with light curtains, wooden partitions and cabinets. This creates a fluid core within the robust structure, which wraps around the interior of the house - like a protective shell.

The four-story houses, each currently occupied by a family, can later be used as shared apartments for the elderly, or individual floors can be separated off as granny apartments. Living is conceived here as the totality of our activities and makes use by the occupants possible even in a constellation not yet defined today in an unwritten future.

-by David Kasparek-

Location Stuttgart, Germany **Use** Housing **Gross floor area** 253m² **Building area** 476m² **Gross floor area** 130m² **Completion** 2022 **Project manager** Marcia Nunes **Photographer** Zooey Braun

코로나19 팬데믹을 지나면서 토지 가격, 건축 비용 및 생활과 일의 균형이 다시금 중요한 가치를 획득한 가운데 슈투트가르트 남쪽의 가파른 경사면에 세워진 쌍둥이 건물은 현재와 미래의 다양한 삶의 형태를 고려한 설계 디자인을 보여준다.

이웃 주택의 처마와 건물의 폭에 맞춘 반 단독 주택으로 디테일한 부분에서 조립식 주택의 형식을 지향했다. 최대 3층 높이까지 경사면을 낮춰야 하는 상황에서 건축 계획 및 시공 비용을 낮추기 위해 심플한 디테일이나 연결부에는 조립용 콘크리트 부품을 활용하였다. 이산화탄소 배출로 인한 재료의 단점이 있지만 다양한 도면 설계에 따라 확장 및 개조가 가능해 장기간 유연하게 사용하는 것으로 이를 보완할 수 있었다. 십자형 평면도에 따르면 내부 공간에 라이트 커튼, 목재 칸막이 및 캐비닛을 다양하게 적용했다. 이것으로 단단한 껍질이 있는 것 같은 견고한 구조의 집 내부에 유동적인 중심을 세우는 것이다.

현재 개별 가족이 거주하고 있는 4층 주택은 이후 노인을 위한 공동 아파트나 층이 분리된 개별 아파트로 활용할 수 있다. 이곳에서는 어떤 생활도 가능하며 거주자로 하여금 아직 도래하지 않은 미래의 다양한 삶의 가능성까지도 향유할 수 있도록 포용한다.

-데이비드 카스파렉-

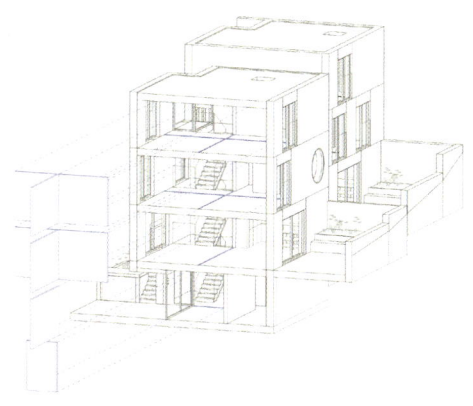

AXONOMETRIC

← Exterior view ↙ Front view

↑ Southeast view   ↙ Partial view

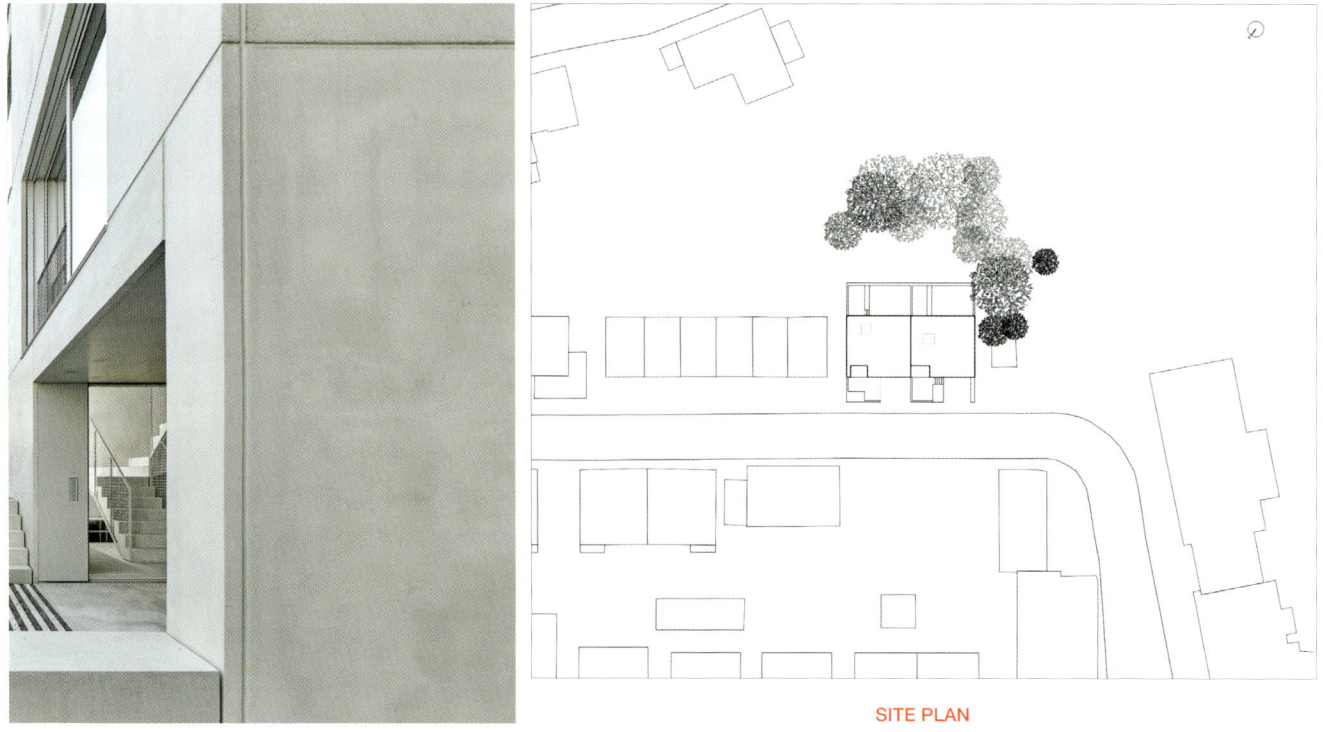

SITE PLAN

← Entrace → Entrace

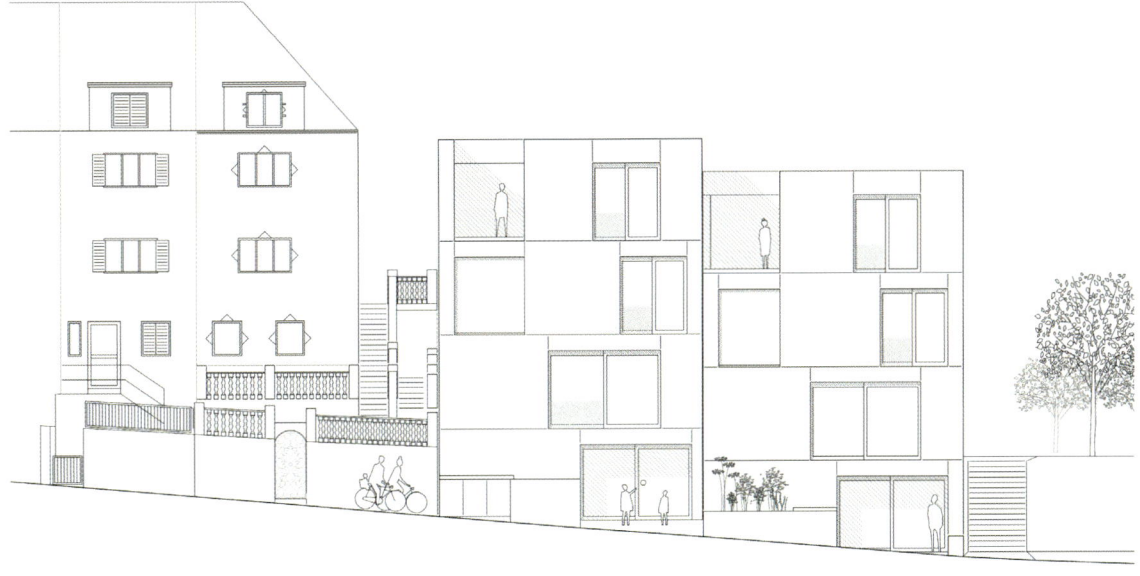

NORTHWEST ELEVATION

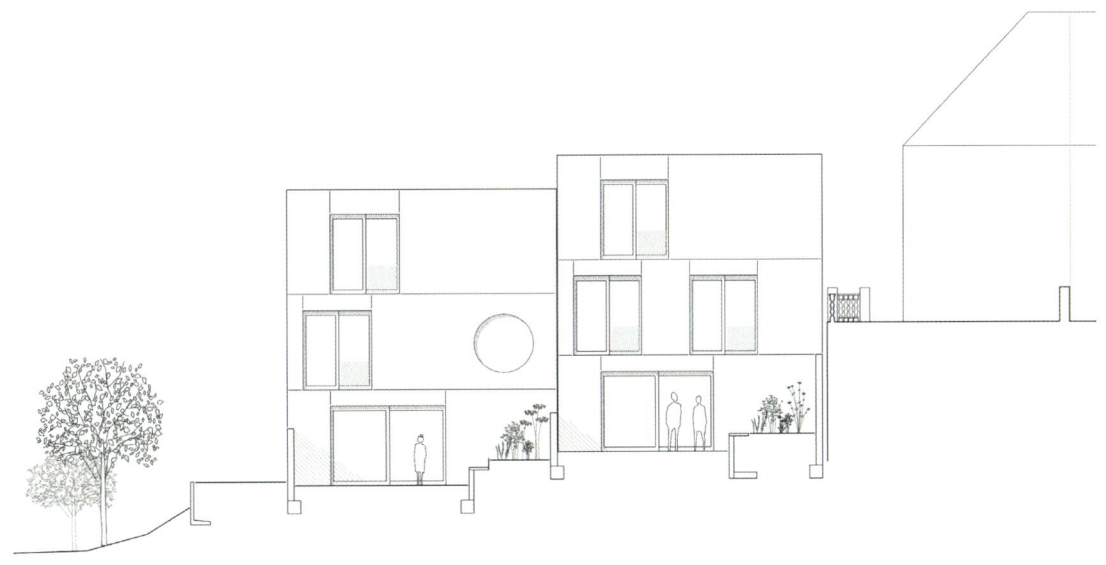

SOUTHEAST ELEVATION

147

1. Terrace

↑ Interior view

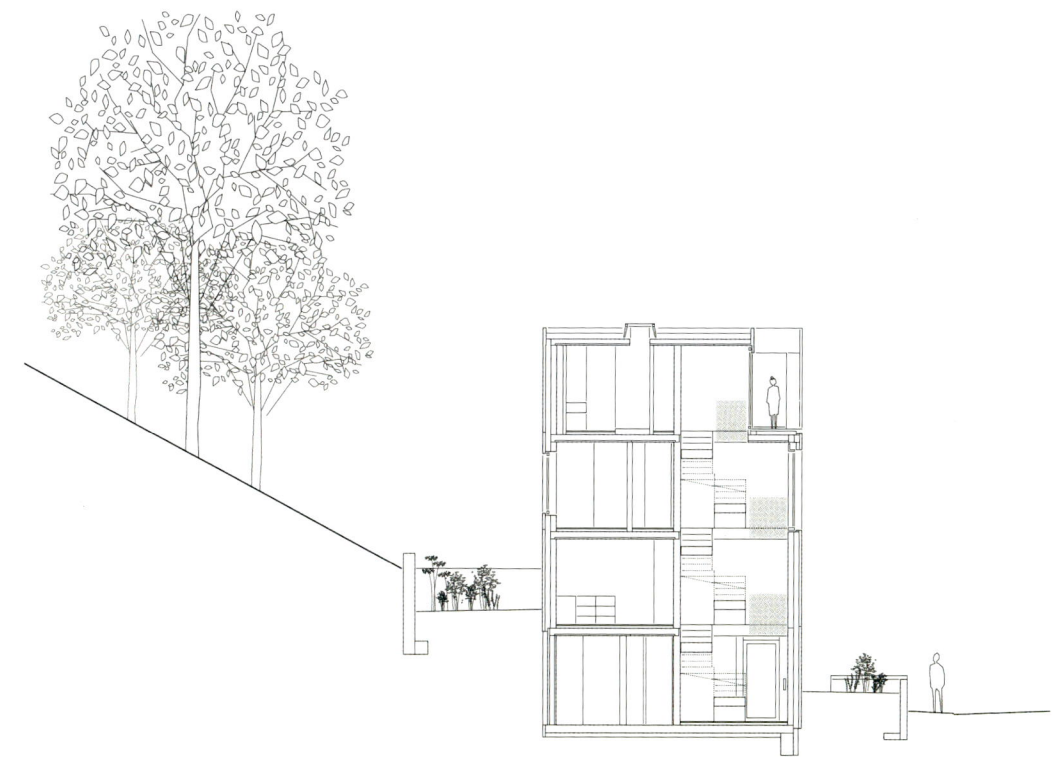

SECTION

↑ Living room ↓ Dining room

← Stairs ↑ Stairs & Study → Hallway

1 ENTRANCE
2 BICYCLE PARKING LOT
3 TOILET
4 LIVING ROOM
5 KITCHEN & DINING ROOM
6 TERRACE
7 BEDROOM
8 BATHROOM
9 STUDY

1ST FLOOR PLAN

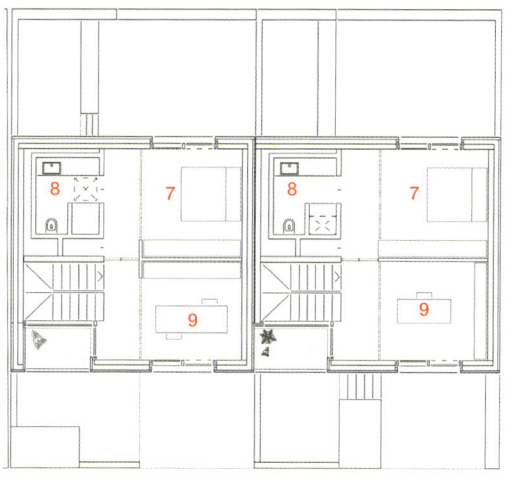

2ND FLOOR PLAN

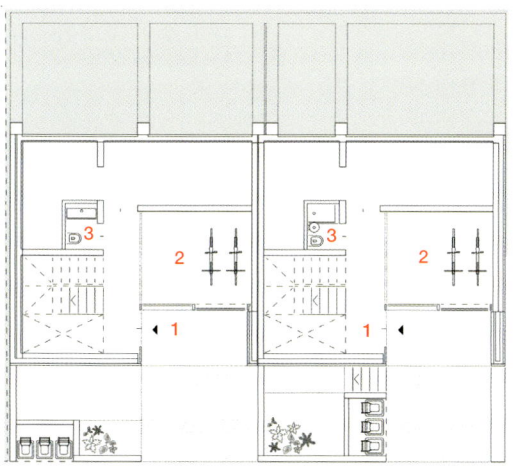

BASEMENT FLOOR PLAN

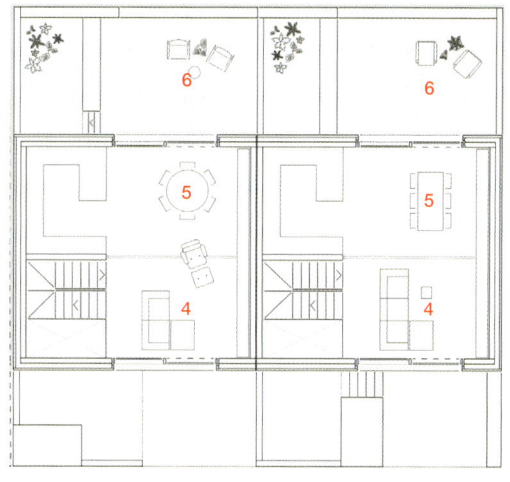

GROUND FLOOR PLAN

비에스피20 하우스
BSP20 HOUSE

ARCHITECT : RAUL SANCHEZ ARCHITECTS / RAUL SANCHEZ

THE REQUIREMENTS WERE TO CONVERT THIS SMALL BUILDING Hlocated in the Borne district of Barcelona, from the end of the 19th century and with 4 floors (but barely 20m² per floor), into a place where to be able to work.

The new floors (3 in total) will be supported by new beams between the dividing walls, which will not touch either of the two facades: towards the main facade, a sheet of glass will separate them from it; and towards the interior facade, the stairwell will be a 4-story void that unites the entire interior and shows the surprising height of such a slender building. Thus, the project is defined, remaining to define the uses of each floor and resolve the technical issues.

The presence of the installations, by ruling out grooves in the walls or small wells from the beginning, take on a special and relevant role inside: 7 stainless steel cylinders run the entire height of the building, conducting all the electrical, ventilation, plumbing, extraction, sanitation, air conditioning and telecommunications installations inside 6 of the cylinders, leaving one of them empty for future needs. These cylinders are not hidden and run through the building through furniture and floors. The rest of the installations are always visible, never built-in, highlighting the roughness of the masonry walls on which they are located, freeing them from new servitudes.

The kitchen is a frosted brass piece of furniture, shiny and with reflections, with a white marble top; the bathroom equipment is paneled with lacquered wood in a slightly cream color, with black and brass details; the 'headboards' of the floors are covered in white microcement; the hydraulic mosaic, microcement and oak floors add warmth and color to the interior; and the lacquered wood ceilings incorporate registers and grids to 'design' these needs.

The structure is all painted white, in search of a certain material abstraction, especially in the development of the spiral staircase, which is developed as a free-standing cylinder that runs the entire height of the building without touching its walls at any time, offering Piranesian views helped by the heterogeneity of the walls and the diversity of points of view.

On the contrary, all the details on the existing walls are direct and raw: the window frames are made with direct mortar, the pre-frames are not concealed, and the structural elements of ties are left unpolished. Above, on top of the stairwell, a skylight introduces a beautiful gradation of light until the lowest strata.

The main facade was rehabilitated following the strict dictates of the heritage department, returning it to an image of the past that it surely never had. Only at the entrance door were we free to invent a front that reproduces the three-dimensional design of the classic hydraulic mosaic (used on the ground floor and much loved by the client) with an exploded view of rhombuses and triangles finished with 3 types of aluminum, which conceals the door (only recognizable by the lock) and abstracts the entrance.

Location Barcelona, Spain **Use** House **Site area** 20m² **Built area** 105m² **Completion** 2022 **Project manager** Raul Sanchez **Architecture** Valentina Barberio **Structure consultant** Diagonal Arquitectura **Engineering** Marés ingenieros **Photographer** José Hevia

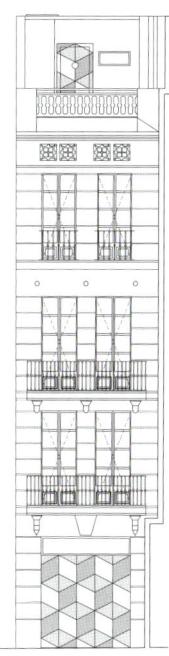
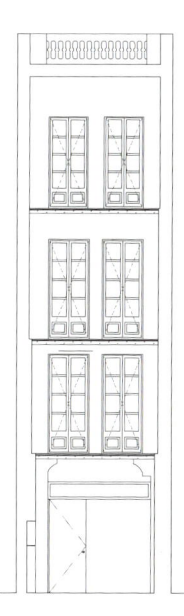

ELEVATION

← Exterior view

East facade view — Front night view

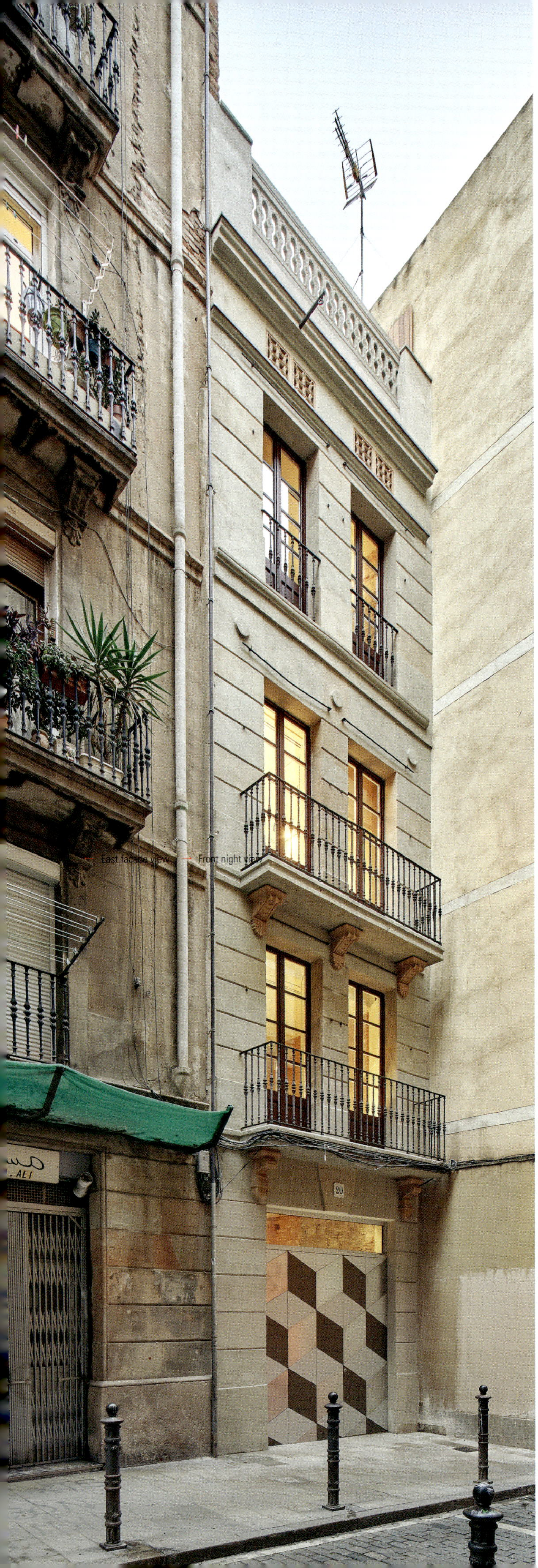

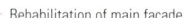

Rehabilitation of main facade

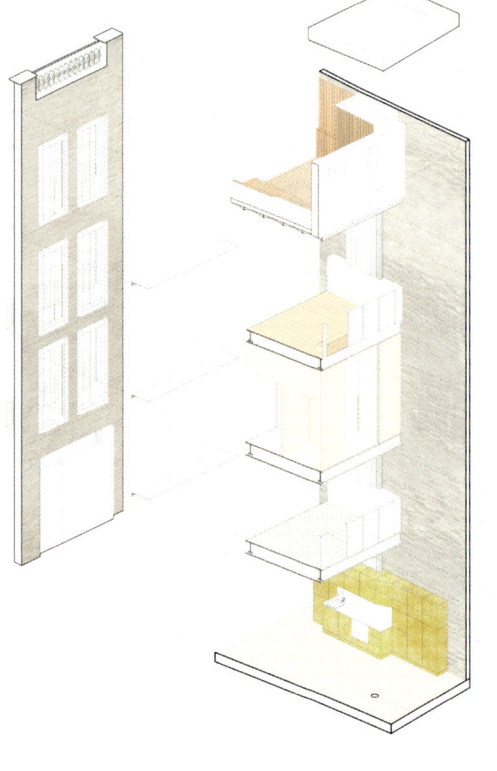

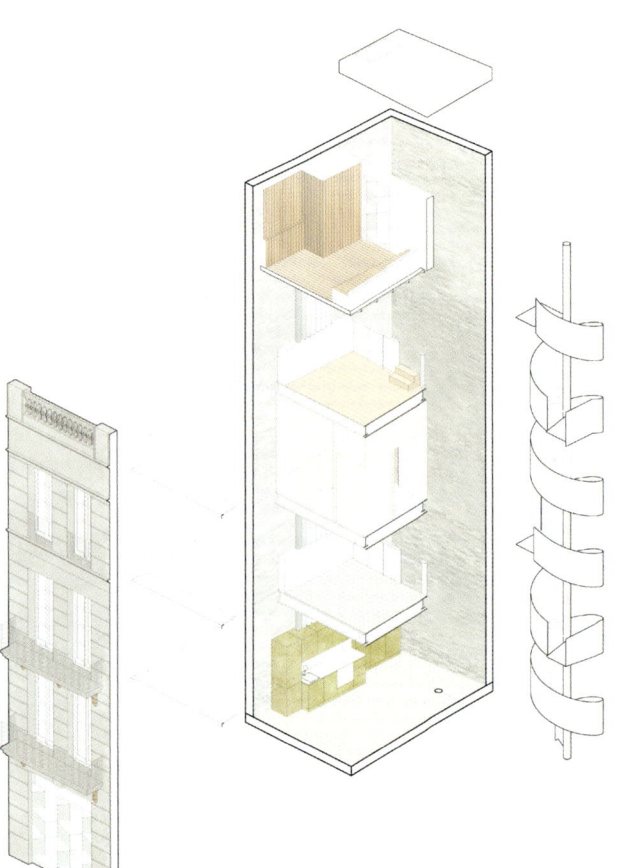

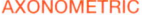

AXONOMETRIC

← Entrance door, three-dimensional design of the classic hydraulic mosaic   → 1st floor stairwell

19세기 말에 바르셀로나 보른 지구에 세워진 작은 규모의 H 건물을 4층(층당 약 20㎡)짜리 업무 빌딩으로 바꾸는 것이 필요했다.

새로운 층(총 3개)은 분할된 벽 사이에 추가되는 기둥으로 지지가 되어 두 외벽 사이에 공간이 생겼다. 외벽으로 향한 유리판으로 분리됐다. 내부 외벽으로는 4층 높이의 계단이 들어서 인테리어를 통합시키면서 슬림한 건물의 높이를 더욱 높여 보이게 했다. 결과적으로 프로젝트의 컨셉은 분명해졌으나, 각 층의 용도와 기술적 문제는 여전히 남아 있었다.

처음부터 벽의 홈이나 작은 통로를 제외하고 내부 용도에 맞는 역할을 하는 설치물은 다음과 같다. 건물 전체에 연결된 스테인리스 스틸 실린더 7개 중 6개의 내부에는 전기, 환기, 배관, 배출, 위생, 에어컨 및 통신을 위해 점유되었으며, 나머지 하나는 이후 필요한 경우를 대비해 비어 있다. 해당 실린더는 숨겨져 있지 않고 가구와 바닥을 통해 건물 내에 설치되어 있다. 나머지 설치물은 내장되어 있지 않고 겉으로 드러나 있으며, 설치된 석조 벽의 거친 면이 고스란히 드러나 개조를 기다리고 있다. 주방에는 은은하게 반짝이며 빛을 반사하는 반투명 황동 가구에 흰 대리석 상판을 더했다. 욕실 가구는 은은한 크림색으로 래커 처리한 목제 패널에 검은색과 황동 장식이 들어가 있다. 바닥의 겉면은 흰색 마이크로 시멘트로 덮고 유압 모자이크와 마이크로 시멘트, 오크 바닥재로 마감해 따뜻한 실내 분위기를 연출했다. 천장도 래커 처리한 목재로 마감해 전체적으로 통일된 느낌을 제공한다.

전체적인 구조는 흰색 색상의 페인트로 마감했고, 소재의 추상미를 강조하는 디자인이 특징인데, 특히 나선형 계단에는 독립형 실린더가 벽이 닿지 않게 건물 전체를 통과하고 있어 벽의 이질성과 관점의 다양성을 포용하는 피라네지적 공간을 구성했다. 반면 기존 벽의 디테일은 소재 본연의 질감을 강조했다. 창틀 일부에는 직접 모르타르를 발라 기존 창틀의 일부가 드러나고 전체적으로 거친 질감이 느껴진다. 계단상부 위에서 비치는 채광은 아래층으로 갈수록 아름다운 그라데이션을 연출한다.

외벽은 문화유산을 관리하는 부서의 엄격한 지시에 따라 복원되었으며, 다시는 돌아갈 수 없는 과거의 모습으로 되돌려졌다. 유일하게 변형할 수 있었던 입구 문에는 마름모꼴과 삼각형으로 이루어진 클래식한 유압 모자이크를 3차원으로 재현해 3가지 알루미늄으로(잠금장치로만 구별되는) 문을 가려 입구를 추상적인 구조물로 만들었다.

SITE PLAN

← The kitchen is a frosted brass piece of furniture   → 1st floor kitchen & entrance door

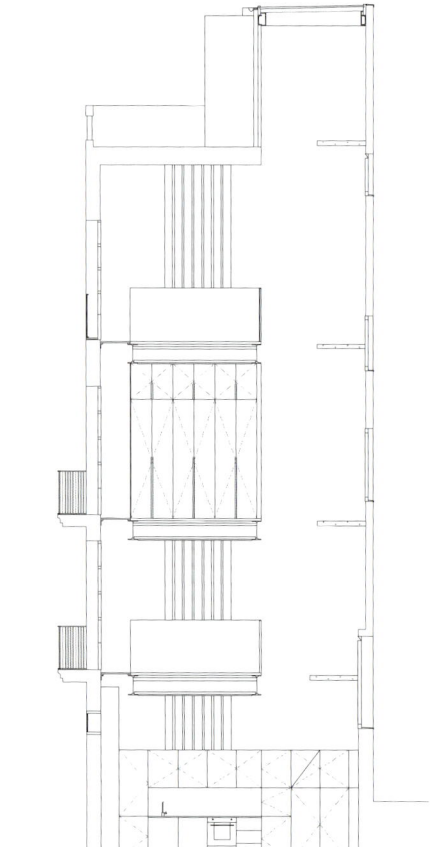
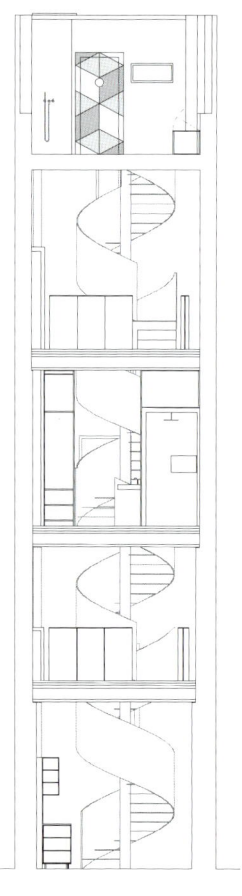
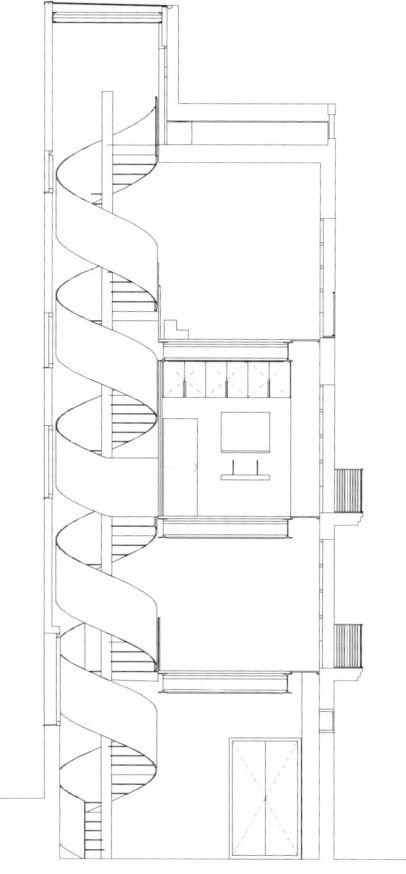

SECTION

↑ 7 stainless steel cylinders entire in the building ↙ The main facade was rehabilitated ↘ 2nd floor stairwell

← 3rd floor bathroom   → 3rd floor bathroom & stair well   ↳ 3rd floor bathroom detail   ↳ Shower room

↑ 4th floor window, rehabilitation   ↙ 4th floor space   ↳ 4th floor stairwell

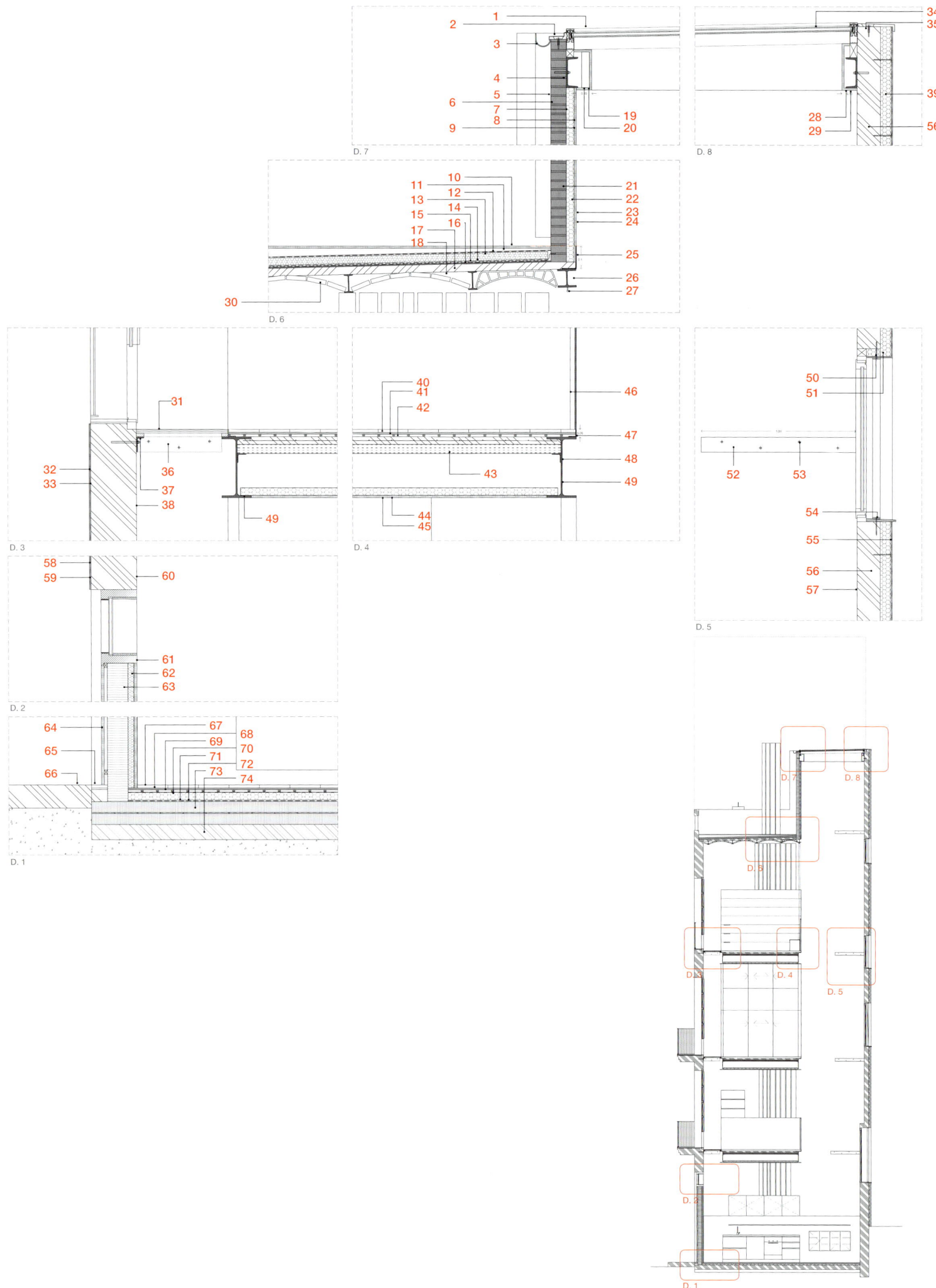

SECTION DETAIL

1. ALUMINUM SKYLIGHT/METAL CARPENTRY TYPE V05 (SEE DRAWING A030)
2. E1mm PRE-LACQUERED STEEL FOLDED SHEET FINISH PLACED WITH MECHANICAL FIXINGS AND SEALING, PROVIDED WITH DRIP
3. STAINLESS STEEL GUTTER
4. UPN 200 BEAM
5. TRADITIONAL PLASTER COATING OF AIR LIME AND LIME PAINTS
6. GERO PERFORATED BRICK ENCLOSURE
7. WALL LINING WITH GALVANIZED STEEL STRUCTURE 8=46mm TYPE PLASTERBOARD + ROCK WOOL FILLING
8. PLASTERBOARD PLATE E 15mm
9. ENAMEL PAINT COATING COLOR TO BE DEFINED BY THE DF
10. CLASS IV TREATED PLANK FLOORING
11. CLASS IV TREATED PINE BATTENS FOR LEVELING.
12. POLYPROPYLENE FELT GEOTEXTILE SHEET PLACED WITHOUT ADHERING.
13. 60mm EXTRUDED POLYSTYRENE (XPS) SHEET INSULATION PLACED WITHOUT ADHERING
14. POLYPROPYLENE FELT GEOTEXTILE SHEET LAID WITHOUT ADHERING, FROM 100 TO 110 g/m²
15. WATERPROOFING WITH A SURFACE DENSITY MEMBRANE OF 6.1 kg/m² FORMED BY SHEET OF LBM MODIFIED BITUMEN, ADHERED AND HOT
16. LEVELING LAYER OF CEMENT MORTAR AND 2cm
17. REINFORCED CONCRETE COMPRESSION LAYER E 5cm
18. EXISTING CERAMIC VAULT FLOOR AND CERAMIC JOIST
19. ENAMEL PAINT COATING COLOR TO BE DEFINED BY THE DF
20. DRAWER WITH GALVANIZED STEEL STRUCTURE TYPE PLASTERBOARD E 46mm AND PLASTERBOARD PLATE E 15mm
21. GERO PERFORATED BRICK ENCLOSURE
22. WALL LINING ITS GALVANIZED STEEL STRUCTURE E=46mm PLASTERBOARD TYPE + ROCK WOOL FILLING
23. PLASTERBOARD PLATE E 15mm
24. ENAMEL PAINT COATING COLOR TO BE DEFINED BY THE DF
25. ANGULAR STEEL PROFILE ROOF SLAB FINISH
26. HEB 120
27. STAINLESS STEEL CURTAIN ROD FIXED TO THE BEAM DIRECTLY
28. ENAMEL PAINT COATING COLOR TO BE DEFINED BY THE DF
29. DRAWER WITH GALVANIZED STEEL STRUCTURE TYPE PLASTERBOARD E 46mm AND PLASTERBOARD PLATE E 15mm
30. RECOVERY OF THE EXISTING CERAMIC VOLTA SLAB, AFTER REMOVING IT DE-COATINGS AND APPLICATION OF COLORLESS FIXATIVE BAMIZ
31. LAMINATED GLASS 10+10+6 NON-SLIP, TRANSPARENT, PLACED ON NEOPRENE
32. TRADITIONAL PLASTER COATING AIR LIME AND LIME PAINTS
33. EXISTING FACTORY ENCLOSURE SOLID BRICK AND 30cm TYPE CE1
34. ALUMINUM SKYLIGHT/METAL CARPENTRY TYPE V05 (SEE DRAWING A030)
35. FINISH OF FOLDED PRE-LACQUERED STEEL PLATE E1mm PLACED - WITH MECHANICAL FIXINGS AND SEALING, PROVIDED WITH GOTERON
36. PLATE 100.8
37. LD 100.50.8
38. BAMIZ FIXATIVE COATING COLORLESS AFTER REMOVAL OF PREVIOUS COATINGSE
39. EXTERNAL THERMAL/ACOUSTIC INSULATION SYSTEM (SATE), COMPOSED OF A 60MM DOUBLE-DENSITY ROCK WOOL PANEL, FINISHED WITH 1MM SILICATE MORTAR, CIRCULAR FLOAT, COLOR TO BE DEFINED BY THE DF
40. HYDRAULIC MOSAIC 20X20X2
41. SELF-LEVELING MORTAR LAYER
42. UNDERFLOOR HEATING SET (SEE MEMORIES)
43. HAIRCOL 59, GALVANIZED, 1mm COLLAPSIBLE SHEET
44. FALSE CEILING FOR INSTALLATIONS
45. LACQUERED WOODEN TOPS WITH HIDDEN FIXINGS, EXPLODED VIEW ACCORDING TO DETAIL
46. 8mm STEEL SHEET RAILING SET, RAL 90 FINISH
47. ANGULAR STEEL PROFILE PAVEMENT FINISH
48. L 60.60.6 WELDED TO IPE
49. IPE 400
50. BARS FIXED TO WALLS WITH COUNTERSUNK HEAD SCREWS
51. THE F1-TYPE FACADE TURNS AROUND AT THE LINTEL AND IT FORMS A DRIP
52. PLATE 100.853
53. CONNECTORS Ø12
54. SILL PRECAST WHITE CONCRETE SILL
55. EXTERNAL THERMAL/ACOUSTIC INSULATION SYSTEM (SATE), COMPOSED OF A 60MM DOUBLE-DENSITY ROCK WOOL PANEL, FINISHED WITH 1mm SILICATE MORTAR, CIRCULAR FLOAT, COLOR TO BE DEFINED BY THE DF
56. EXISTING ENCLOSURE OF SOLID BRICK MASONRY TYPE C3
57. COLORLESS FIXATIVE BAMIZ COATING AFTER REMOVAL OF PREVIOUS COATINGS
58. TRADITIONAL PLASTER COATING AIR LIME AND LIME PAINTS
59. EXISTING ENCLOSURE OF SOLID BRICK FACTORY AND 30cm TYPE CE1
60. COLORLESS FIXATIVE BAMIZ COATING AFTER REMOVAL OF PREVIOUS COATINGS
61. RAL 9003 MICROCEMENT COATING
62. CLADDING ENCLOSURE GALVANIZED STEEL SUBSTRUCTURE E=46mm PLASTERBOARD TYPE + ROCK WOOL FILLING
63. BRICK FACTORY WALL ENCLOSURE GERO PLASTER FINISH
64. ALUMINUM SHEET CLADDING, SMOOTH-RAW-VARNISH FINISH AS IT UNFOLDS
65. VARNISHED BRASS DOOR SILL PIECE
66. REINFORCED CONCRETE SCREED FLOATED
67. HYDRAULIC MOSAIC 20X20X2
68. SELF-LEVELING MORTAR LAYER
69. UNDERFLOOR HEATING SET (SEE MEMORIES)
70. ANTI-PUNCTURE SEPARATOR LAYER CHOVA GEOFIM PP 100-12 OR SIMILAR
71. THERMAL INSULATION CHOVAFOAM 250 S40 E-40mm OR SIMILAR
72. MOISTURE VAPOR BARRIER/SEPARATOR LAYER
73. CONCRETE SCREED HA 25 8-15cm REINFORCED ME 26 15X15cm
74. CLEAN CONCRETE >10cm UNTIL FIRM

↑ Rehabilitation facade view

↑ Roof top   ↵ Roof top   ↳ Stairwell

← Overall view of the stairwell   → Stairwell

1 ENTRANCE
2 KITCHEN
3 STORAGE FURNITURE
4 STAIRWELL
5 WINDOW
6 BALCONY
7 BATH ROOM
8 TOILET
9 SHOWER ROOM
10 ROOM
11 ROOF TOP

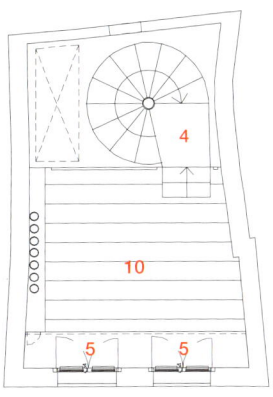

4TH FLOOR PLAN

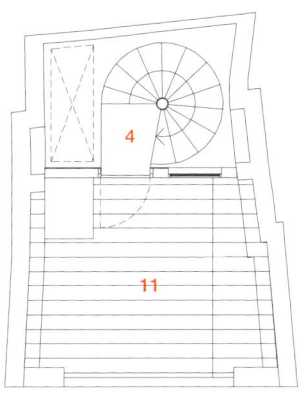

ROOF TOP PLAN

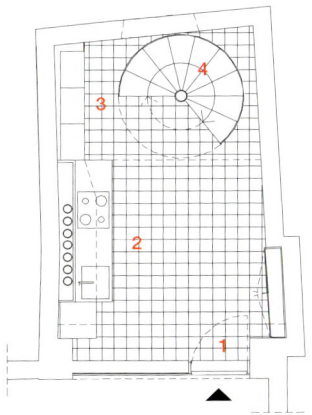

1ST FLOOR PLAN

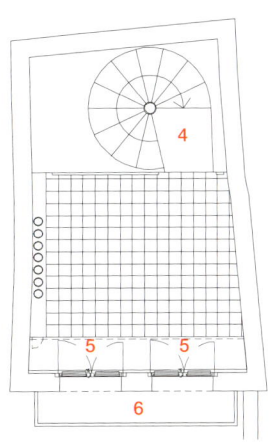

2ND FLOOR PLAN

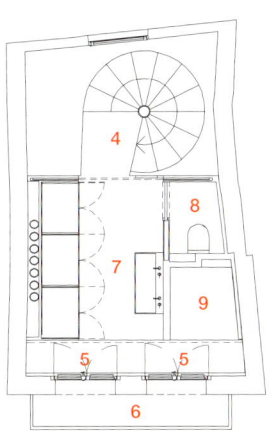

3RD FLOOR PLAN

티비디 사옥
TBD OFFICE

ARCHITECT : Aaaaaa / TRIET LE, HO NGOC NHUNG

TBD IS AN OFFICE FOR OUR PARTNER, a general contractor specializing in interior and fit out. It sits on a rental plot of land whose context strongly influence the building's construction system and its spatial experience that is set to expire in the near future.

Ho Chi Minh is a migration destination, the most populous, and the fastest growing. Yet the city is set to experience the severe effect of sea level rise as the challenging issue of climate change is becoming more prominent. The building implements practical features that are commonly found in tropical vernacular architecture, especially those around the Mekong Delta region. It sits on stilts to avoid potential flooding, facilitate natural ventilation, allow water to be naturally absorbed deep into the ground, and refill the aquifer underneath. As common as it is in the rural area, this practice is overlooked as the city develop its urban fabric. As a developer of their own product, the design addresses the full cycle of the building which offers options to either move, or upcycle parts of the building to new constructions, or dismantle for scraps to recoup part of their initial investment when the lease ends. TBD is as much an architectural challenge for our contractor partner to go beyond their line of work and expand their market, as it is for us to address the fast-changing future of the city.

Foreign as it looks, the building offers a familiar experience. By borrowing the surrounding landscape that is yet to be developed, the building spatial perception embodied the openness that position itself in an in-between state, betwixt inside and outside, and embracing the nature of tropical vernacular. This effect is further enhanced by the use of monochromatic interior that blur the hard building line and its corner.

Through this project, we hope to present an alternative to our industry in responding to the brief and the unspoken condition that is affecting the city's urban fabric. At the same time, offering a feasible shift to employ and modernize our local knowledge, recognizing the intrinsic connection between rural and urban area as they are developing simultaneously.

Location D2, Thu Duc, Ho Chi Minh, Vietnam **Use** Office **Site area** 240m² **Completion** 2022 **Develop & Construct** Thai Binh Duong Constructio **Photographer** Hiroyuki Oki

티비디 프로젝트는 인테리어 및 설비를 전문으로 하는 종합 건설 업체인 파트너사의 사무실이다. 이 사무실 공간은 가까운 미래에 만료될 예정이어서 건물의 구조에 큰 영향을 미칠 임대 부지에 세워져 있다.

호치민은 인구가 가장 많고 빠르게 증가하고 있는 이주 도시이다. 하지만 더욱더 뚜렷해지는 기후 변화 문제로 인해 해수면 상승이라는 심각한 영향을 받게 되었다. 특히 이 건축물은 메콩강 삼각주에 일반적으로 세워진 열대성 주택 양식으로 실용적인 기능을 구현한다. 잠재적인 홍수를 방지하고 자연적인 환기를 용이하게 하며 물이 자연스럽게 땅 깊숙이 흡수되고 아래의 함수층을 다시 채우기 위해 기둥 위에 세워졌다. 농촌 지역에서 발견되는 흔한 현상으로, 이러한 관행이 도시가 발전되는 동안 간과되어 왔다. 건물을 설계하는 개발자는 임대 기간이 종료될 때 설계를 통해 건물을 새로 짓거나 일부를 개조하거나 초기 투자금 일부를 회수하기 위해 건물을 해체하는 방식으로 건물의 수명을 관리한다. 우리가 급변하는 도시의 미래에 대응해야 하는 것처럼 파트너사인 티비디에게는 업무의 경계와 시장의 영역을 확장해야 하는 건축적인 과제가 주어졌다.

이국적인 분위기의 이 건물은 동시에 친근한 느낌을 준다. 아직 개발되지 않은 주변의 경관을 활용하여 건물의 공간이 내외부 중간 상태에 머물며 열대 지방 특유의 분위기를 포용할 수 있는 개방적 구조를 구현하였다. 건물의 견고한 라인과 모서리를 흐릿한 단색으로 마감하여 이 효과를 더욱 강화하였다.

이 프로젝트를 통해 도시의 구조에 은근히 영향을 미치는 환경에 대응하여 우리 산업에 대한 대안을 제시하고자 한다. 또한 농촌과 도시 지역이 동시에 발전하면서 본질적으로 연결된다는 것을 인식하여 현지 자원을 활용하고 현대화할 수 있는 실현 가능한 변화를 제공한다.

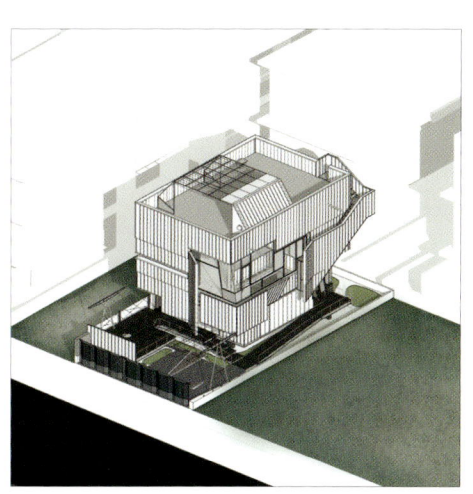

CONTEXT DIAGRAM

← Front facade

← Exterior view   → Exterior view

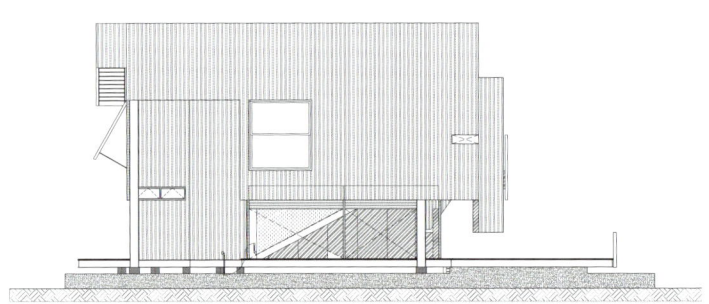

LEFT ELEVATION

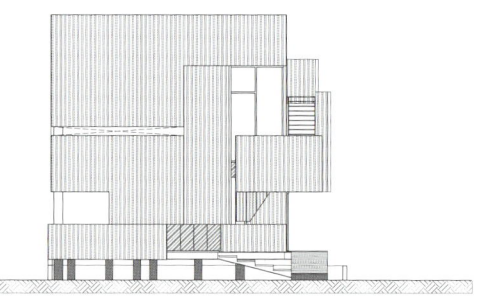

FRONT ELEVATION

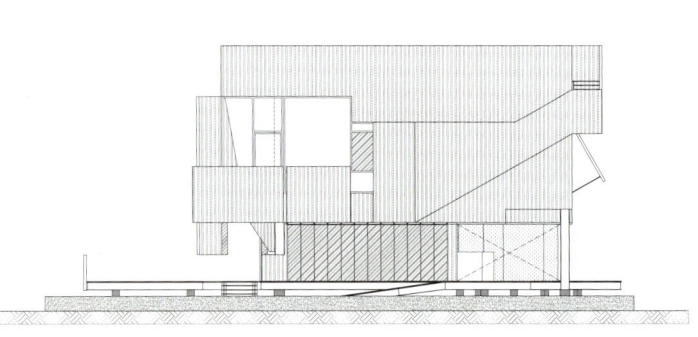

RIGHT ELEVATION

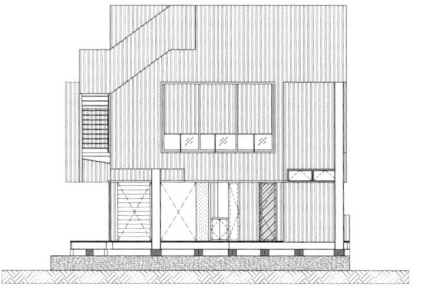

BACK ELEVATION

↑ Corner view   ↙ Panoramic view   ↳ Distant view

167

↑ View on the ground floor   ↙ Entrance

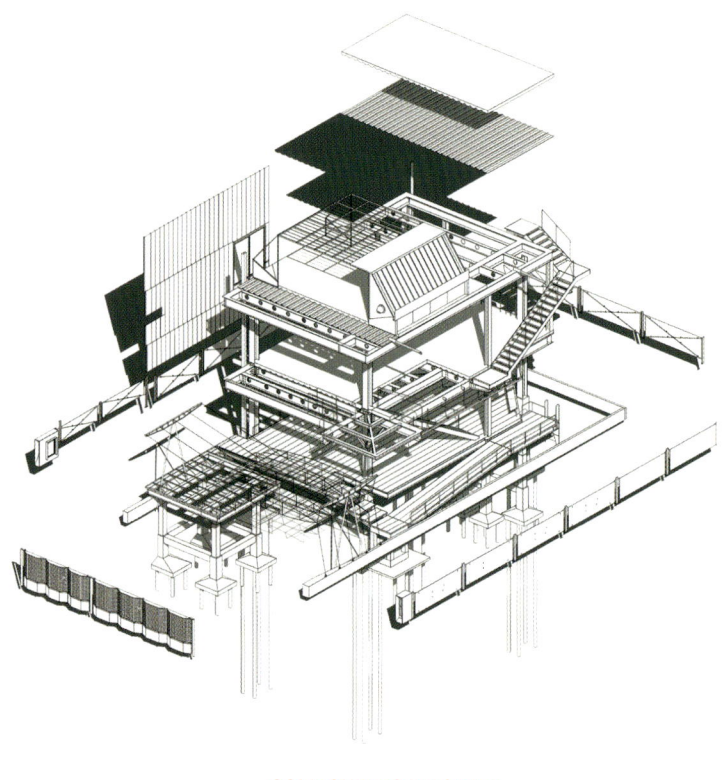

COMPONENTS DIAGRAM

← Stairs → Interior view

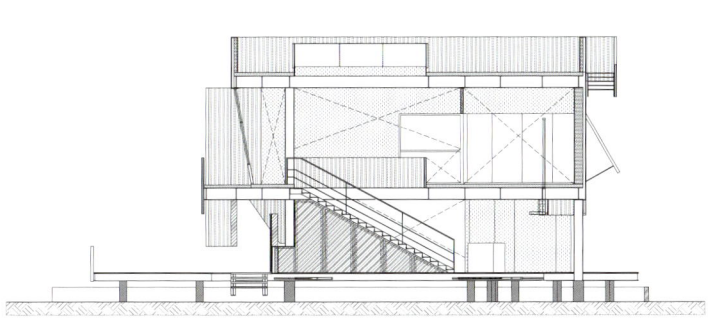
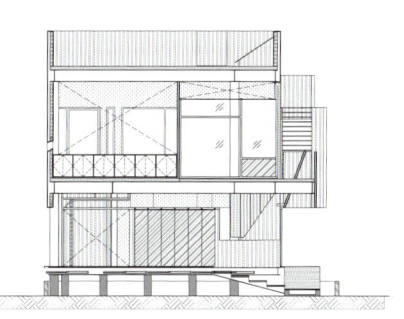
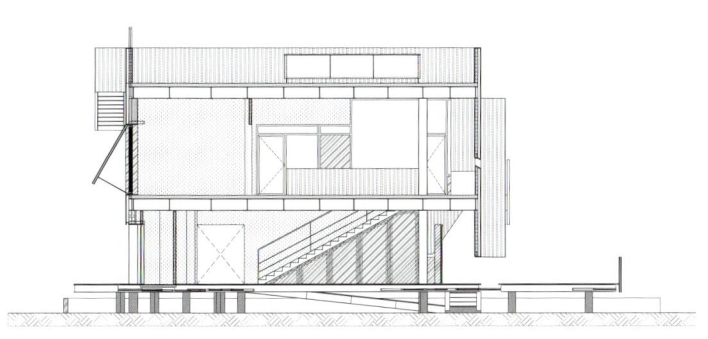
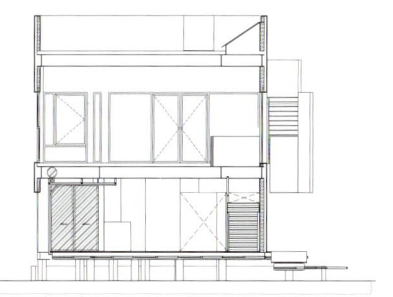

SECTION

↑ Interior view

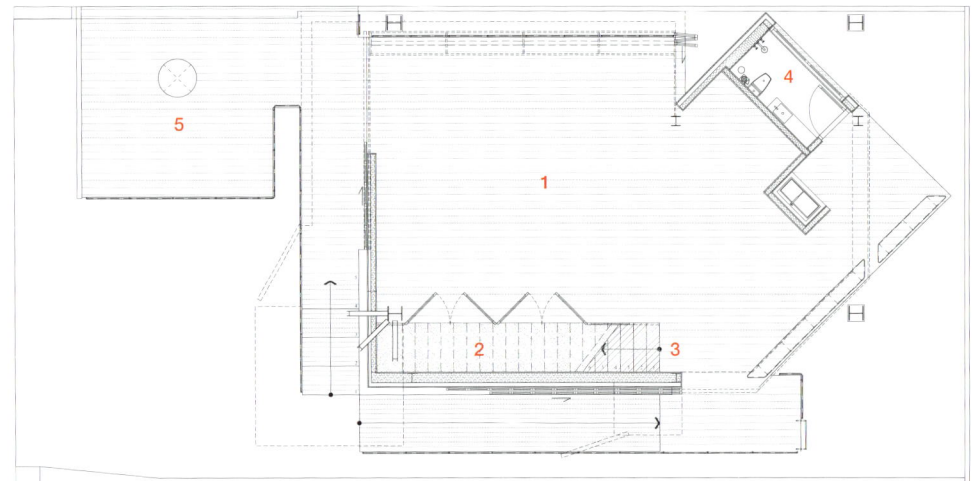

GROUND FLOOR PLAN

1 SHOWROOM
2 STORAGE
3 STAIRS
4 RESTROOM
5 TERRACE
6 OFFICE
7 MEETING ROOM
8 ROOFTOP GARDEN

↑ Office ↵ Office ↳ Office

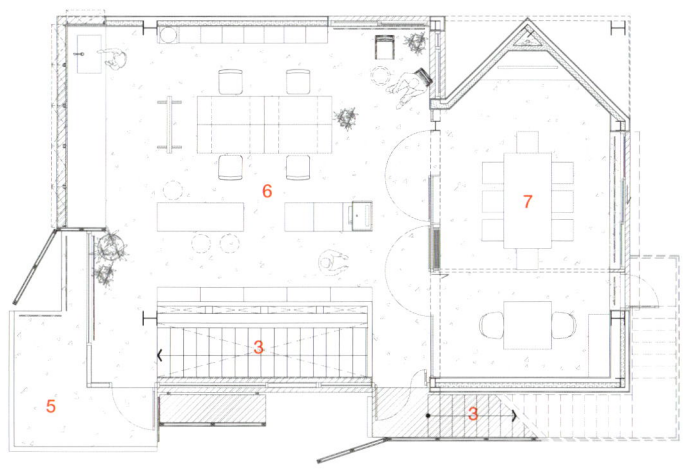

1ST FLOOR PLAN   ROOFTOP FLOOR PLAN

더즌 도어스
DOZEN DOORS

ARCHITECT : GON ARCHITECTS / GONZALO PARDO

DOZEN DOORS IS THE RESULT OF THE TRANSFORMATION OF AN EARLY 21ST CENTURY single-family house with five floors above ground and a basement, located in the multicultural neighborhood of Tetuán in Madrid, into a co-living where 12 university students.

The project is an urban interior structure, dedicated to accommodating people united by similar life positions that favor coexistence, while at the same time leaving space for the sphere of individuality. To this end, the general organization of the built volume, conceived both in plan and in section, is articulated around a staircase strategically placed in the center of the building, which, while establishing the vertical circulations of the co-living, distributes and organizes horizontally both the common and private spaces.

From the collective point of view, there is a diversity of indoor and outdoor communal spaces distributed throughout the building, where different types of equipment are shared. From the communal kitchen, dining room or living room where to watch the FIFA World Cup or the latest music video trends on the ground floor, or the games room in the basement, where the laundry room is also located, to the south-facing terraces where to sunbathe overlooking the Madrid skyline, on the fourth and fifth floors, passing through each of the landings on each level, where, next to the lockers, more unexpected interactions can take place. In short, these public spaces are projected as indeterminate places that facilitate encounters and conversation.

The private rooms, unlike the common spaces, are projected as variations on the same type, all the same and at the same time different. They are the places where the personal worlds of each student can evolve. They contain, in 10~12m², all the necessary elements for hygiene (bathroom), rest (bed) and work (study area). The use of color, as in the case of the study area and the bathroom, chosen according to each orientation, serves to differentiate these areas from the rest of the room and hierarchize them spatially.

Location Madrid, Spain **Use** Co-Living **Site area** 915m² **Built area** 325m² **Gross floor area** 312m² **Completion** 2022 **Project manager** Carol Pierina Linares **Design team** Cristina Ramírez, Laura Argüeso, María Cecilia Cordero, Iván Rando, Kostís Toulgaridis, Celia Urbano **Contractor** EDIAR SLU **Photographer** Imagen Subliminal (Miguel de Guzmán + Rocío Romero)

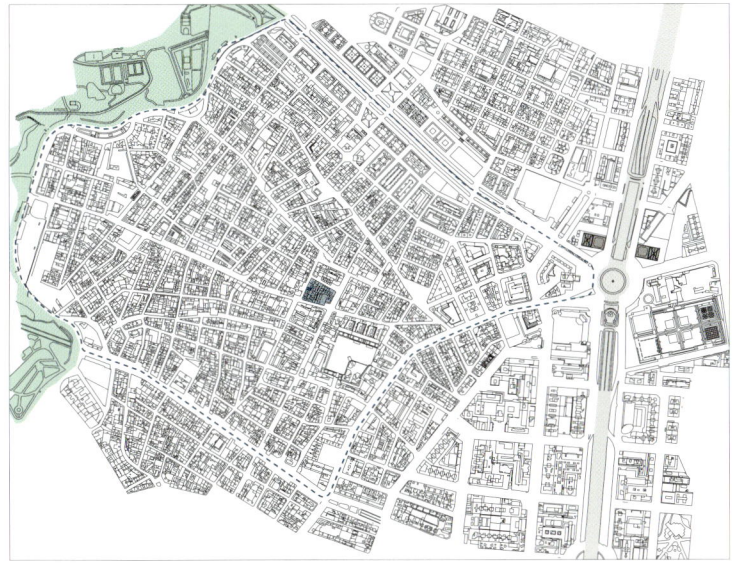

SITE PLAN

← Exterior & entrance view

↑ Entrance & communal kitchen   ↙ 1st floor, communal living room   → Communal kitchen, dining room

SECTION

ISOMETRIC VIEW OF FLOOR

ISOMETRIC VIEW OF ROOM 1

ISOMETRIC VIEW OF ROOM 2

다문화가 공존하는 마드리드의 테투안 지구에 위치하는 지상 5층, 지하 1층 규모의 21세기 초 단독 주택을 12명의 대학생이 거주하는 공동생활 주택으로 바꾸면서 12개의 문을 구성하게 되었다.

이 건물은 공동생활을 원하는 비슷한 처지인 사람들이 동시에 독립적으로 삶을 영위할 수 있도록 내부 구조를 현대적으로 변경하는 프로젝트이다. 이를 위해 설계 및 단면도 상의 기본적인 건축 구조는 건물 중앙에 전략적으로 배치된 계단 주위에 수직적 공동생활 공간을 구성하고 공용 공간과 개인 공간은 수평적으로 구성했다.

공동으로 생활이 가능하도록 다양한 실내 및 실외 공용 공간을 건물 전체에 배치하고 다양한 장비를 공유할 수 있게 설치했다. 1층의 공용 주방부터 식당, 피파 월드컵 게임을 시청하거나 최신 음악을 들을 수 있는 거실, 지하의 게임룸과 세탁실, 마드리드의 스카이라인이 펼쳐진 곳에서 일광욕을 즐길 수 있는 4, 5층의 남향 테라스는 물론 우연한 만남이 일어날 수 있는 각 층의 락커 옆 계단참이 특징이다. 요컨대, 이런 공동 공간에서는 우연한 만남도, 대화도 한층 편안하게 이뤄진다.

개인 공간은 공용 공간과 달리 같은 유형의 공간이지만 동일한 시간대라도 개인에 따라 변주되는 곳이다. 이곳에서 각 학생의 본인 세계를 확장할 수 있다. 10~12㎡의 해당 공간에는 위생(화장실), 휴식(침대), 및 작업(학습 공간)이 적절하게 구비되어 있다. 학습 공간과 화장실에는 용도에 맞는 색상이 적용되어 휴식 공간과 계층적으로 분리된다.

← 1st floor bed & study area   → Bathroom   ↙ 1st floor terrace   ↓ Stair system   ↘ Stair system

↑ 2nd floor bathroom, bed and study area ← Study area ↓ Bathroom → bed

↑ 4th floor terrace   ↙ Terrace entrance   ↓ Roof top stair   ↳ Roof top sunbed

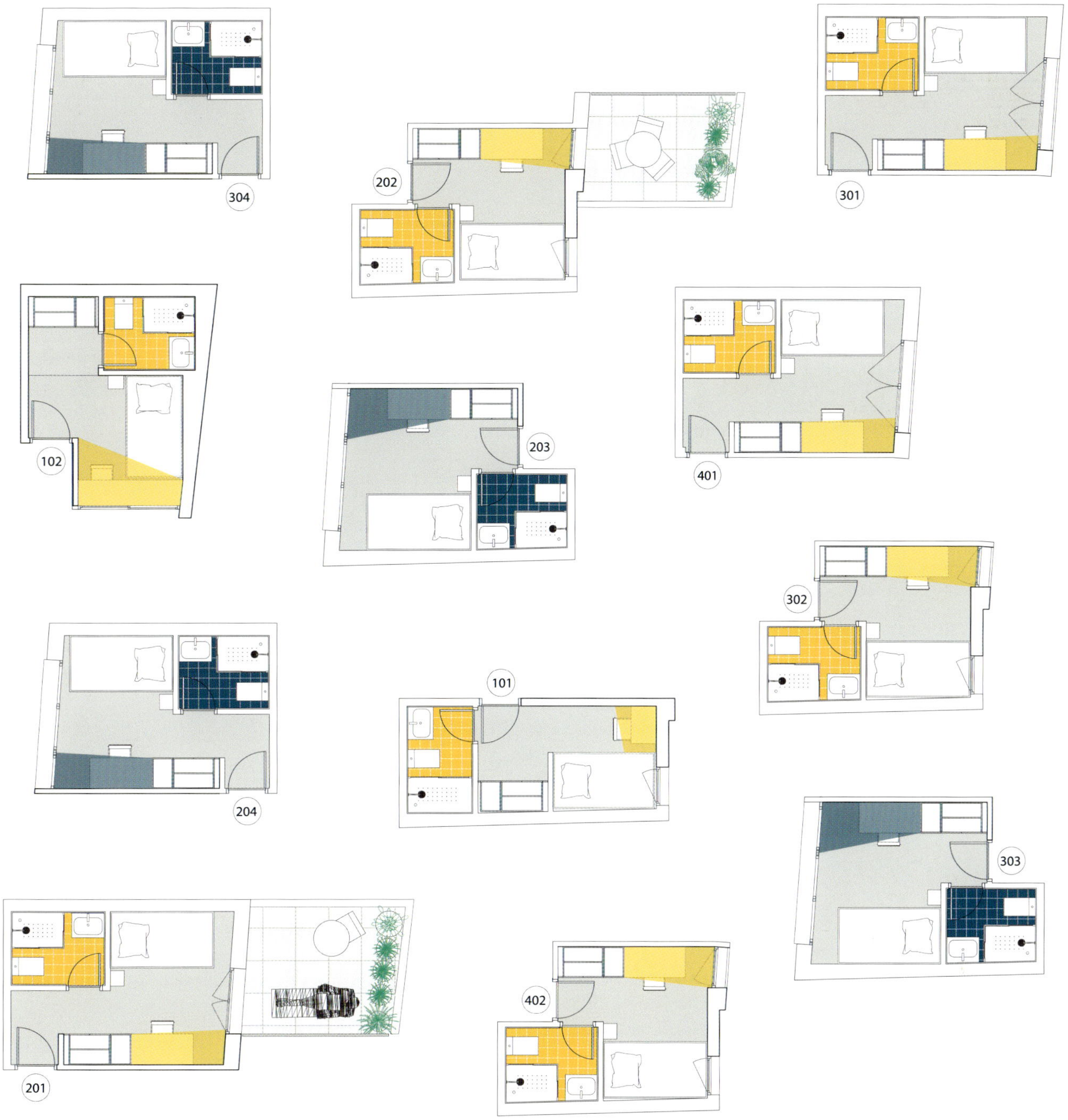

ROOMS TYPE

| 1 | ENTRANCE | 3 | DINING ROOM | 5 | ROOM | 7 | TERRACE |
| 2 | KITCHEN | 4 | LIVING ROOM | 6 | STAIR | 8 | BATHROOM |

1ST FLOOR PLAN

2ND FLOOR PLAN

179

타이베이 공연예술센터

TAIPEI PERFORMING ARTS CENTER

ARCHITECT : OMA / REM KOOLHAAS, DAVID GIANOTTEN

LOCATED AT TAIPEI'S SHILIN NIGHT MARKET MARKED BY ITS VIBRANT STREET CULTURE, Taipei Performing Arts Center is architecture in limbo : specific yet flexible, undisrupted yet public, iconic without being conceived as such. Three theaters plugged into a central cube allow performing spaces to be coupled for new theatrical possibilities. The cube is lifted off the ground for a Public Loop to extend the street life of Taipei into the theater. New internal possibilities and connections of the theater generate different relationships between producers, spectators, and the public, also a critical mass that works as a fresh, intelligent icon.

The central cube consolidates the stages, back stages, support spaces of the three theaters, and the public spaces for spectators into a single and efficient whole. The theaters can be modified or merged for unsuspected scenarios and uses. The spherical 800-seat Globe Playhouse, with an inner and an outer shell, resembles a planet docking against the cube. Intersection between the inner shell and the cube forms a unique proscenium for experimentation with stage framing. Between the two layers of shells is the circulation space that brings visitors to the auditorium. The Grand Theater, slightly asymmetrical in shape and defying the standard shoebox design, is a 1500-seat theater space for different performing arts genres. Opposite to it and on the same level is the 800-seat Blue Box for the most experimental performances. When coupled, the two theaters become the Super Theater-a massive space with factory quality that can accommodate productions that are otherwise only possible in found spaces. New possibilities of theater configurations and stage settings inspire productions in unimagined and spontaneous forms.

The general public. with or without a ticket.is invited into the theater through a Public Loop, which runs through the theater's infrastructure and spaces of production that are typically hidden. Portal windows along the Public Loop allow visitors to look at the performances inside and technical spaces in between the theaters. Different than typical performance centers that have a front and a back side, Taipei Performing Arts Center has multiple faces defined by the theaters protruding above ground. With opaque facades, these theaters appear as mysterious elements against the animated and illuminated central cube clad in corrugated glass. A landscaped plaza beneath the compact theater is an additional stage for the public to gather, in this dense and vibrant part of Taipei.

Location Shilin District, Taipei **Use** Theater **Area** 58,658m² **Completion** 2022 **Clients** Authority-in-Charge: Taipei City Government; Executive Departments: Department of Cultural Affairs, Department of Rapid Transit Systems (First District Project Office), Public Works Department (New Construction Office) **Partners in Charge** Rem Koolhaas and David Gianotten

Design Development Phase (2009 ~ 2013) : Project Architects Ibrahim Elhayawam, Adam Frampton **Team :** Yannis Chan, Hin-Yeung Cheung, Jim Dodson, Inge Goudsmit, Alasdair Graham, Vincent Kersten, Chiaju Lin, Vivien Liu, Kai Sun Luk, Kevin Mak, Slobodan Radoman, Roberto Requejo, Saul Smeding, Elaine Tsui, Viviano Villarreal-Bueron, Casey Wang, Leonie Wenz

Design development phase (2009 ~ 2013) : Project Director Chiaju Lin **Associates** Paolo Caracini, Inge Goudsmit, Daan Ooievaar **Team** Vincent Kersten, Han Kuo, Kevin Mak, Chang-An Liao, with Yannis Chan, Hin-Yeung Cheung, Meng-Fu Kuo, Nien Lee, Nicole Tsai **Photographer** OMA by Chris Stowers, Shephotoerd Co. Photography for OMA, Iwan Baan, Jeffrey Cheng, Frans Parthesius

MODELING

SITE PLAN

Exterior east view   Exterior west view

↑ Overall view of the building ↙ East side view ↘ West side night view

↑ Overall view of the building

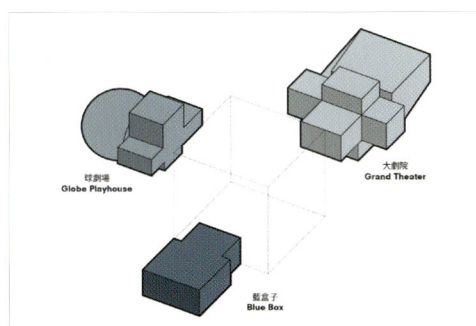 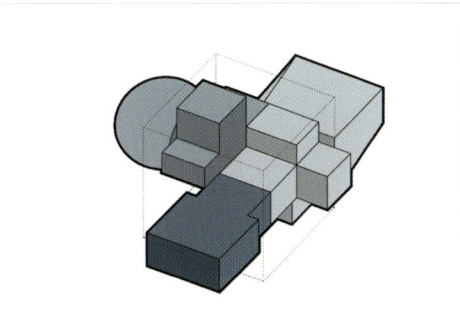 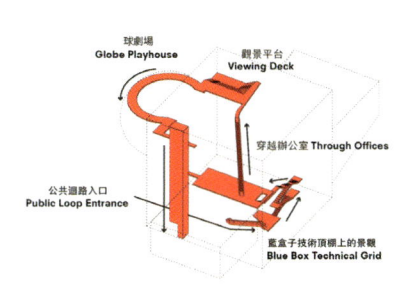

CONCEPT DIAGRAM

활기찬 스린 야시장에 위치한 타이베이 공연예술센터는 사람들의 발길이 뜸해진 곳이다. 특정한 목적을 가졌으면서도 갇히지 않고 대중에게 열려 있는 유연한 곳이지만 큰 상징성이 없는 곳이었다. 중앙의 큐브 공간에 들어선 3개의 극장을 서로 연결하여 새로운 공연을 시도할 수 있게 되었다. 큐브 공간을 지면에서 들어 올려 타이베이의 거리에서 극장으로 자연스럽게 이어 지는 퍼블릭 루프 역할을 한다. 극장 내부의 잠재력과 연결성은 제작자와 관객, 대중 간의 다양한 관계성을 생성하며 신선하고 스마트한 상징으로 작동되는 임계 질량이 되기도 한다.

중앙의 큐브 공간은 3개 극장의 무대와 백스테이지, 공연 준비 공간, 관중을 위한 공공 공간을 하나의 효율적인 공간으로 통합했다. 예상치 못한 시나리오나 활용 목적에 따라 이 극장들을 개조하거나 통합할 수 있다. 800석 규모의 글로브 플레이하우스는 내외부에 쉘이 있는 볼 형태의 구조로 마치 큐브에 행성이 내려앉은 듯한 모습이다. 내부 쉘과 큐브가 교차하는 지점은 독특한 프로시니엄을 만들어 실험적인 무대 구상이 가능하다. 두 층의 쉘 사이에는 방문객을 강당으로 안내하는 순환 공간이 있다. 일반적인 직사각형의 형태가 아니라 약간 비대칭을 이루는 대극장은 1,500석의 규모로, 다양한 공연 예술이 가능하다. 같은 층 맞은편에는 실험적인 공연에 적합한 800석 규모의 블루 박스가 있습니다. 이 두 개의 극장을 통합하면 공장처럼 거대한 공간에서 색다른 연출 효과를 낼 수 있는 슈퍼 극장이 된다. 극장 구성 및 공연 연출의 잠재력은 즉흥적으로 상상치 못한 새로운 연출을 끌어낸다.

티켓 소지 여부와 상관없이 일반 대중은 보통 숨겨져 있는 극장의 인프라와 연출 공간을 연결 하는 퍼블릭 루프를 통해 극장으로 초대된다. 퍼블릭 루프를 따라 구성된 창을 통해 방문객들은 내부 공연 및 극장 사이 기술 공간들을 볼 수 있다. 타이베이 공연예술센터는 앞 뒷면이 있는 일반 공연장과 달리 지상 위로 들어 올려진 극장을 통해 한층 입체적인 효과를 연출한다. 극장의 불투명한 파사드는 골이 진 유리로 둘러싸여 존재감을 드러내는 중앙의 큐브 공간에 신비한 분위기를 더해준다. 소형 극장 아래 조성된 조경 광장은 활기차고 밀집된 타이베이에서 사람들이 모이게 될 또 다른 무대를 제공한다.

↑ Street view

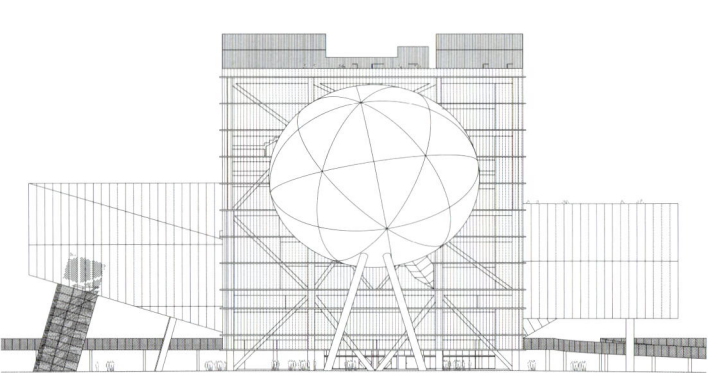

WEST ELEVATION

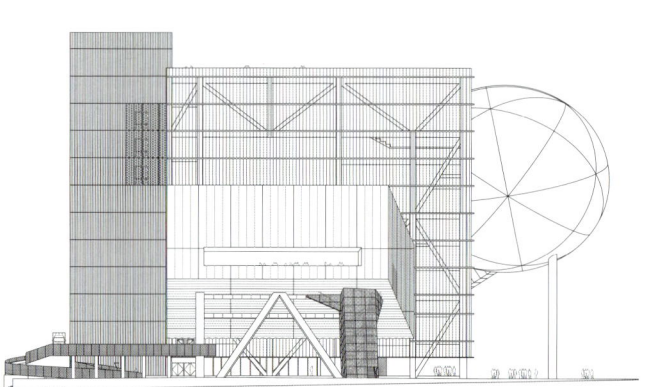

NORTH ELEVATION

EAST ELEVATION

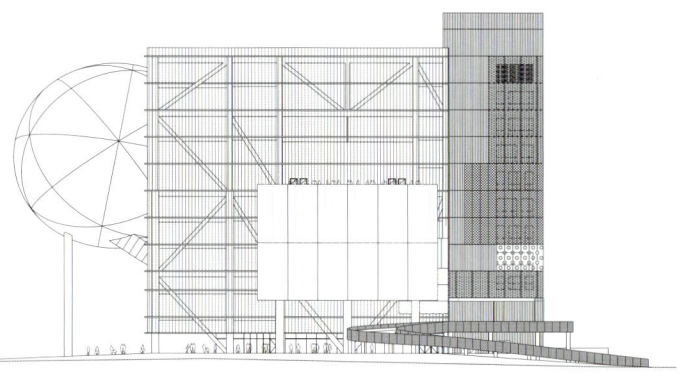

SOUTH ELEVATION

→ Main entrance

SECTIONAL PERSPECTIVE 1

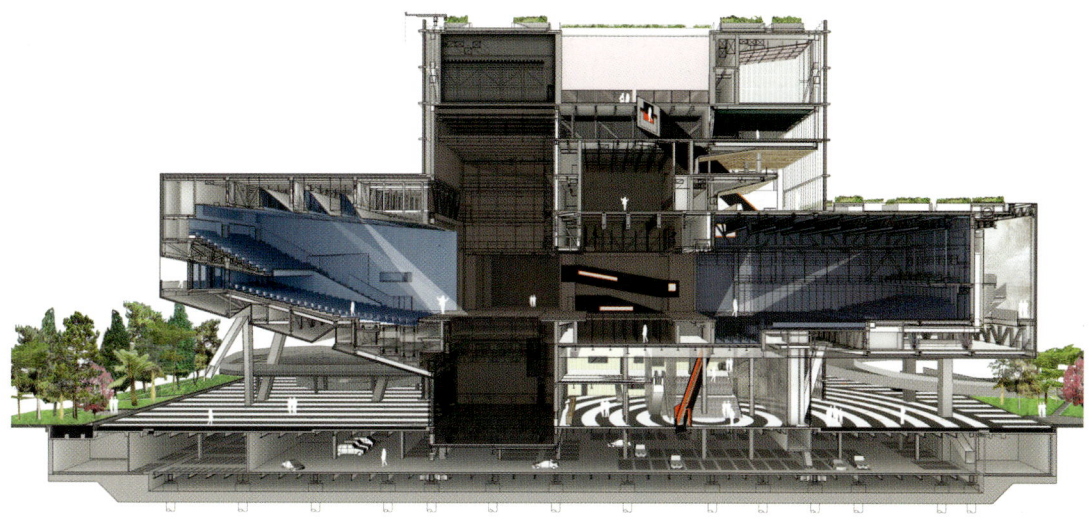

SECTIONAL PERSPECTIVE 2

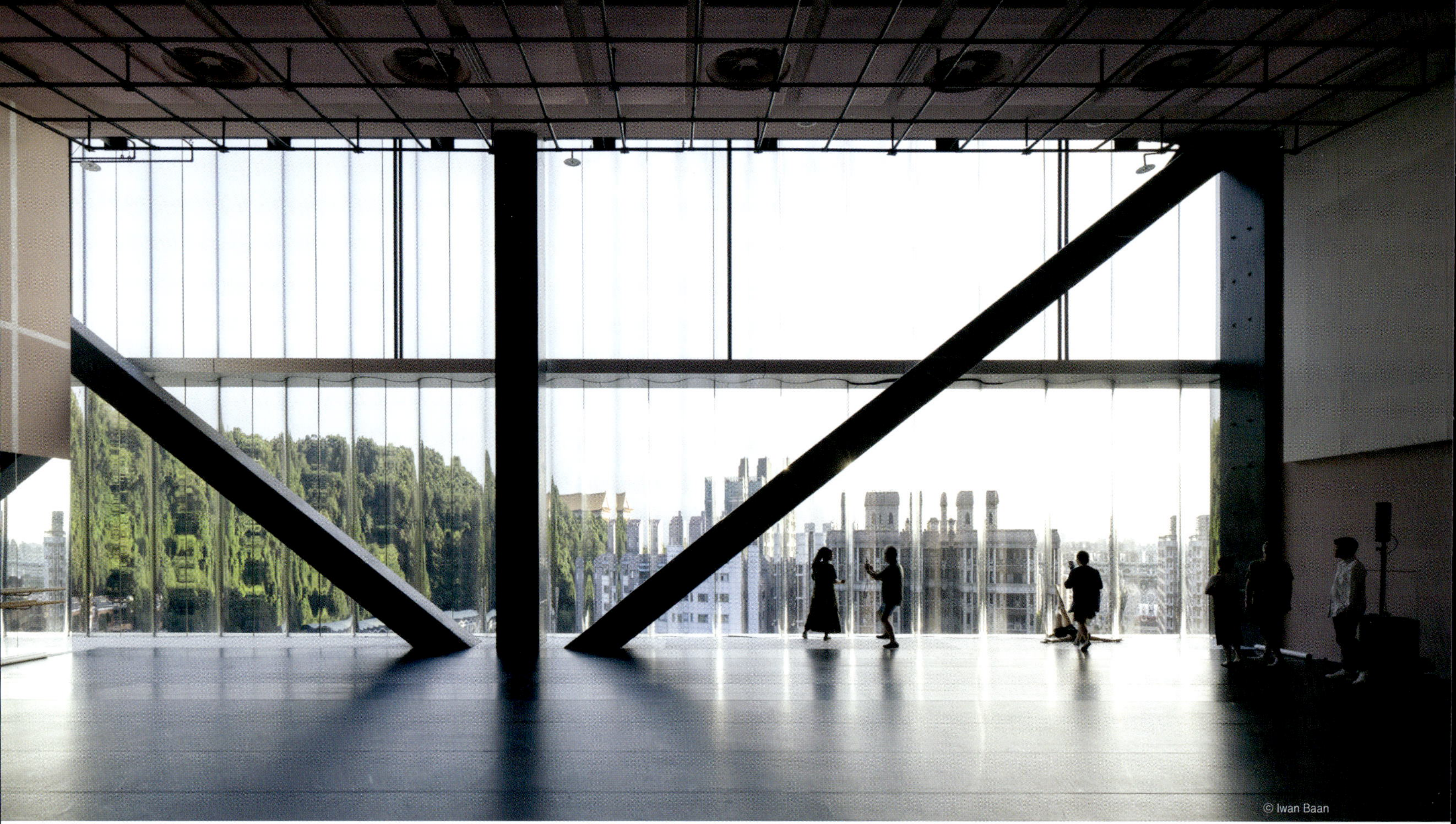

↑ Rehearsal studio ↖ Viewing platform ↱ Auditorium ↵ VIP prefunction bar ↳ Stage

#		#		#		#	
1	PARKING	8	FOYER	15	COURTYARD	22	CONTRAL ROOM
2	PLANT ROOM	9	LARGE REHEARDAL STUDIO	16	CORRIDOR	23	LIGHTING POSITIONS
3	COMMEON LOBBY	10	CAFE	17	PUMP ROOM	24	MEETING ROOM
4	MAIN STAGE	11	TERRACE	18	PUPPET THEATER	25	INFORMATION CENTER
5	REAR STAGE	12	ARCHIVING	19	STAIR	26	LIFT MACHINE ROOM
6	AUDITORTUM	13	PROGRAMMING OFFICE	20	FIRE FIGHT LOBBY	27	FRONT DESK
7	STAGE / PARTERRE LEVEL	14	SMALL REHEARDAL STUDIO	21	SIDE STAGE	28	MEDIUM REHAERSAL STUDIO

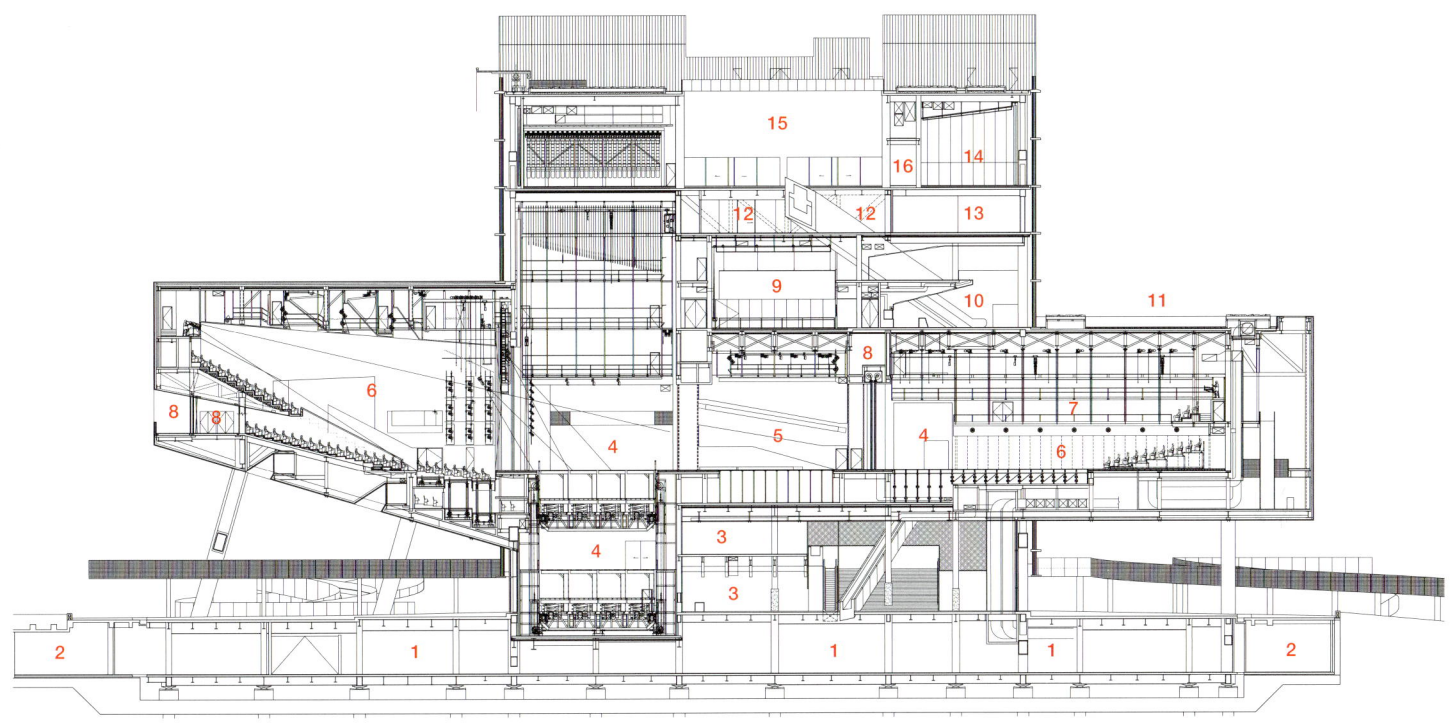

SECTION - A

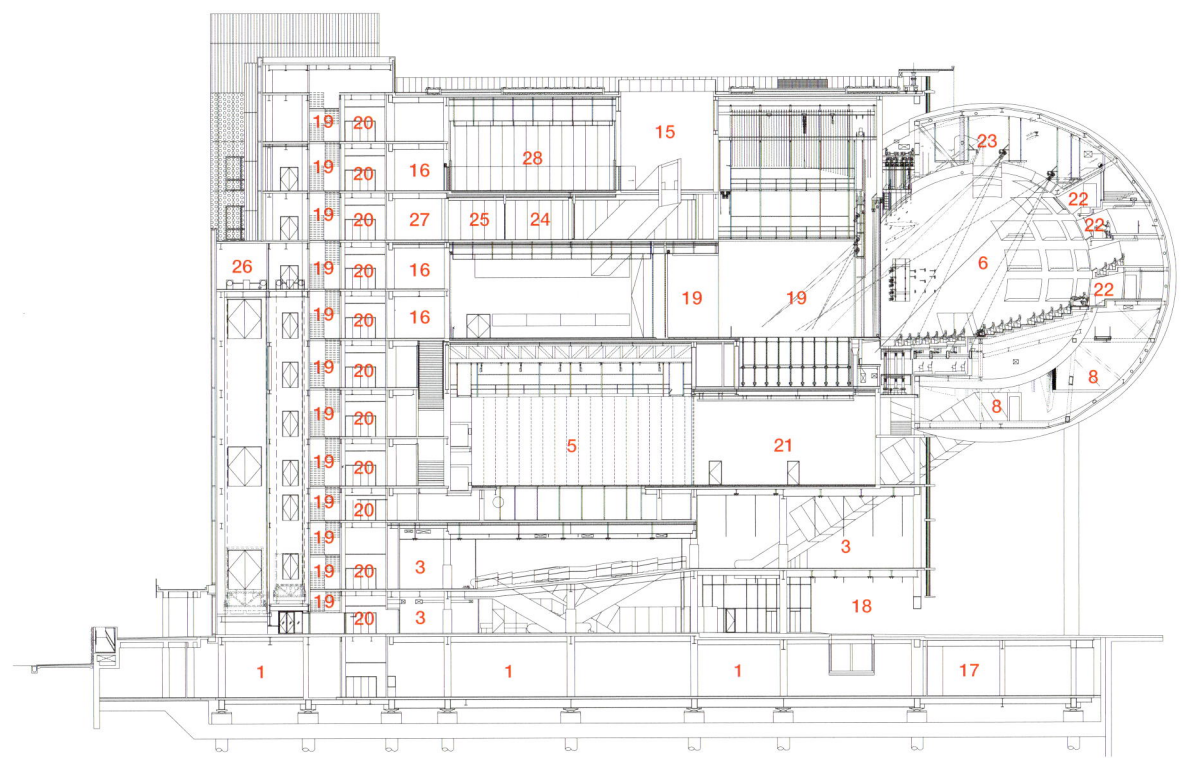

SECTION - B

187

#		#		#		#	
1	PARKING	10	TOILET	19	STAIR & ACCESS	28	CORRIDOR
2	LOBBY PLANT ROOM	11	CIRCULATION	20	KITCHEN / PANTRY	29	SPECIAL SAFETY STAIRCASE
3	COOLING STORAGE	12	SIDE STAGE	21	FRONT DESK	30	GREEN ROOM
4	COMMON LOBBY	13	CIMMER ROOM	22	SMALL MEETING AREA	31	OFFICE
5	SOUVENIR SHOP	14	COSTUME STORAGE	23	RELAX AREA	32	ARTISTC DIRECTORS OFFICE
6	BAR	15	GREEN AREA	24	TAXI DROP-OFF	33	DOCUMENT CONSERVATION ROOM
7	COSTUME STORAGE	16	TERRACE	25	BACK STAGE FRONT DESK	34	INFORMATION & RESEARCH EQUIPMENT
8	LAUNDRY	17	WORKING SPACE	26	LOADING AREA	35	SMALL REHEARSAL STUDIO
9	DRESSING ROOM	18	SEWING AREA	27	AUDITORIUM	36	COOLING TOWER

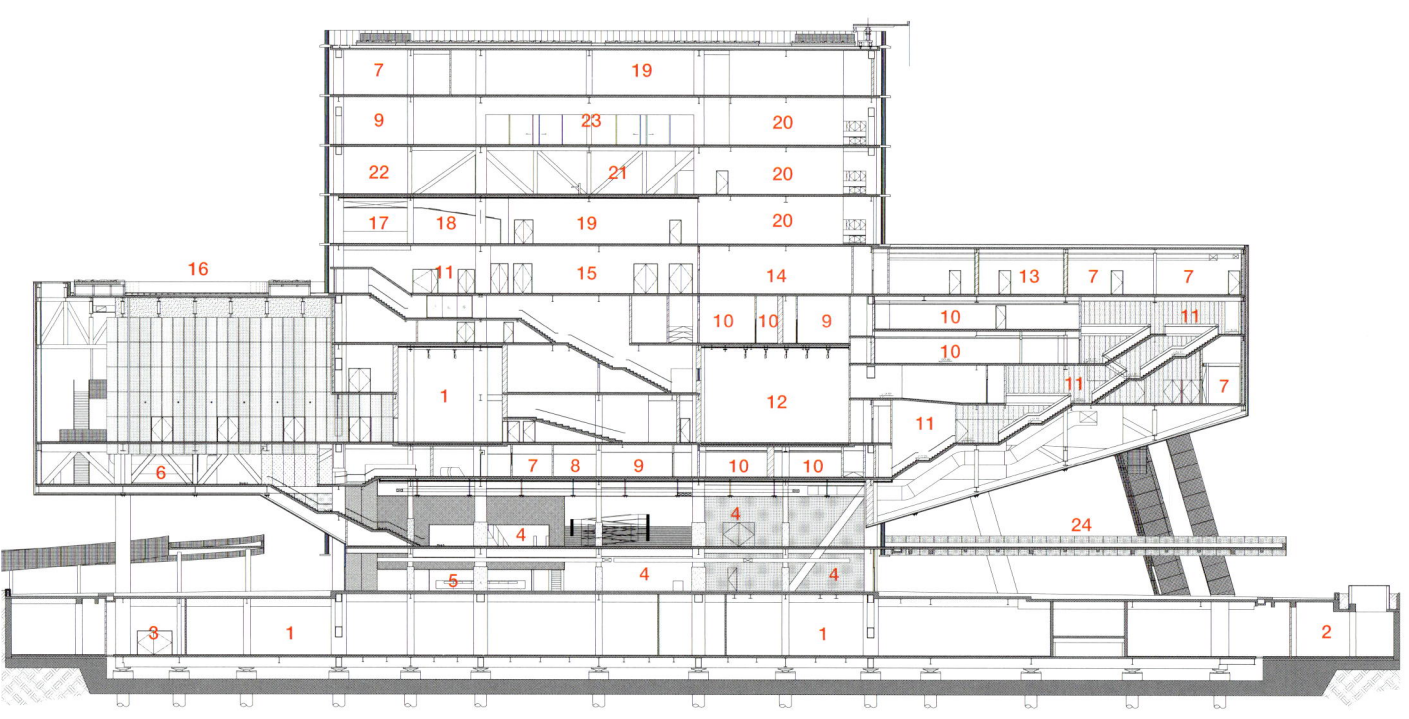

SECTION - C

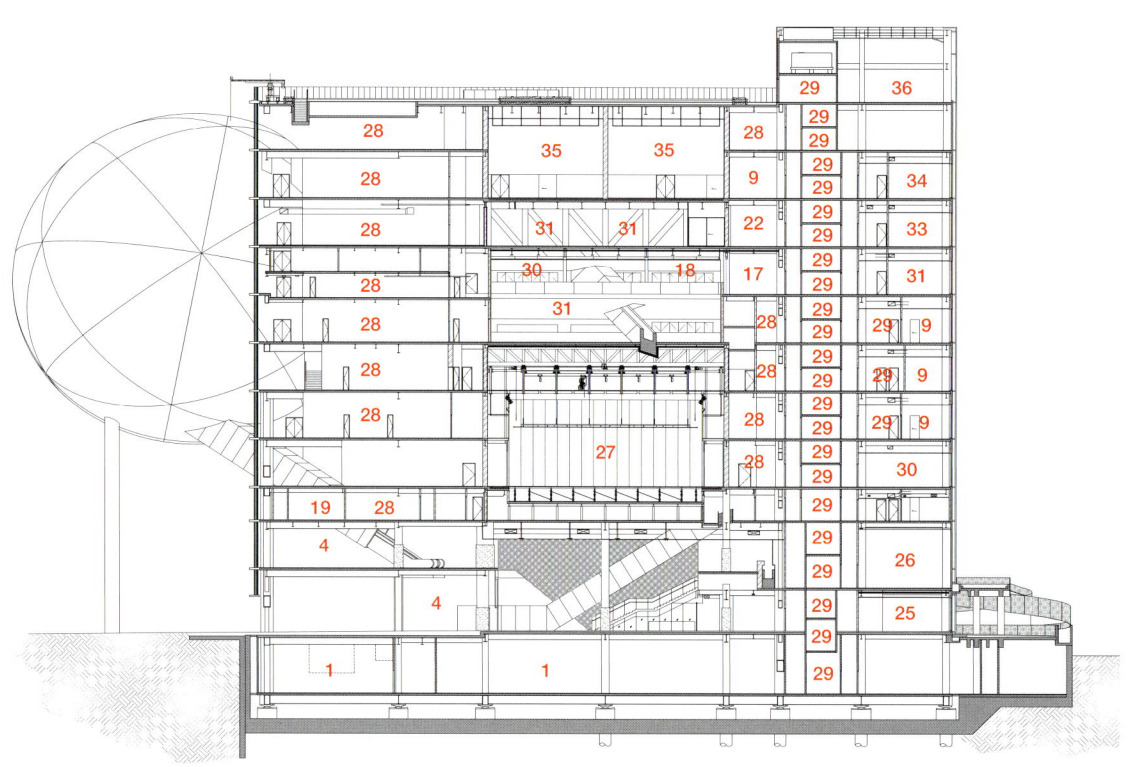

SECTION - D

#		#		#		#	
1	PARKING	9	DRESSING ROOM	17	INFORMATION REDERCH ROOM	25	OUTDOOR VIP BALCONY
2	STAIRS	10	CORRIDOR	18	TERRACE	26	MUSICAL LOUNGE
3	SMOKE LOBBY	11	SIDE STAGE	19	COOLING TOWER	27	UNDER STAGE
4	HVAC	12	STORAGE	20	COMMON LOBBY	28	STAGE
5	STAFF LOUNGE	13	DOCUMENT CONDERVATION ROOM	21	VIP ENTRACE	29	DIMMER ROOM
6	TOILET	14	MULTIMEDIA ROOM	22	ASSEMBLY	30	VIEWING PLATEFORM
7	LOADING AREA	15	ELECTRICAL CONTRAL ROOM	23	INTERIOR EGRESSS STAIR		
8	REPAIR ROOM	16	PLANT ROOM	24	VIP PREFUMCTION BAR		

SECTION - E

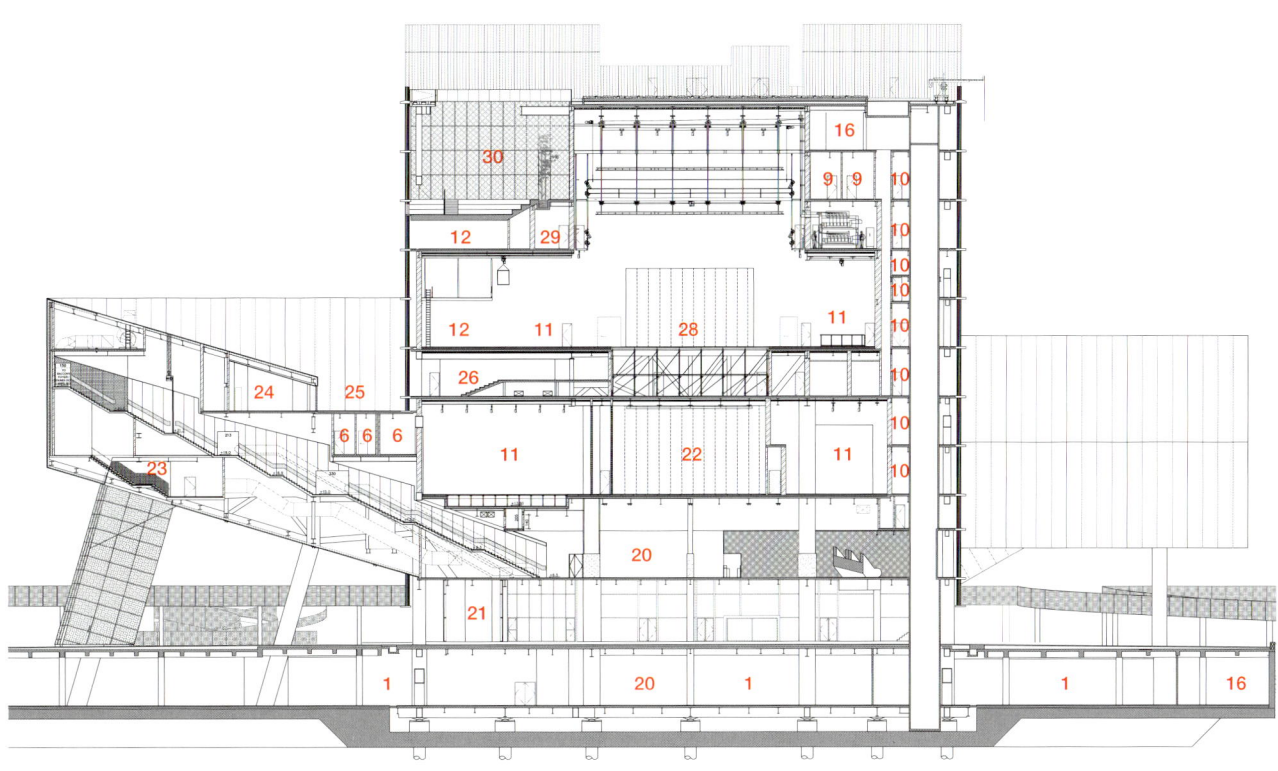

SECTION - F

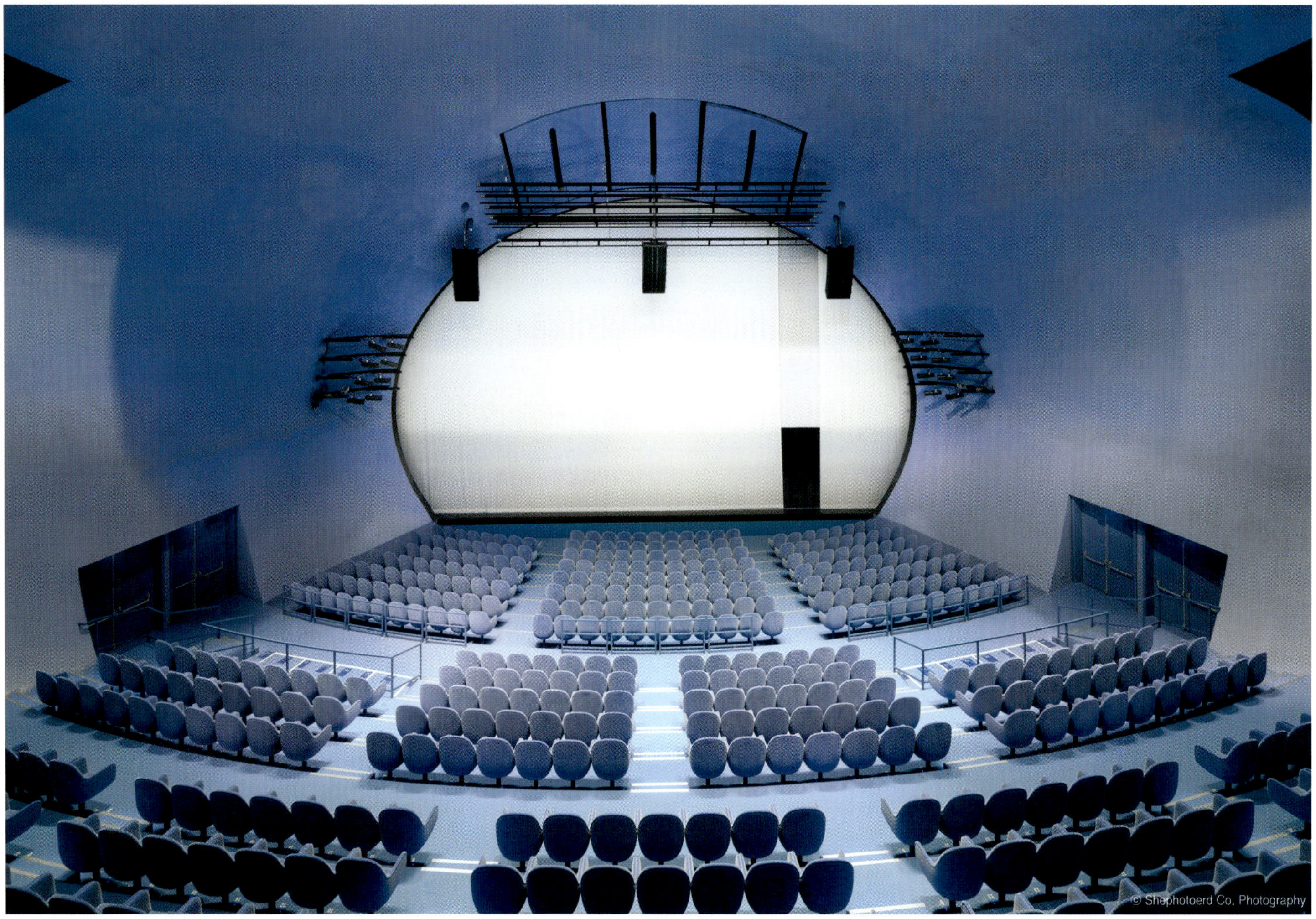

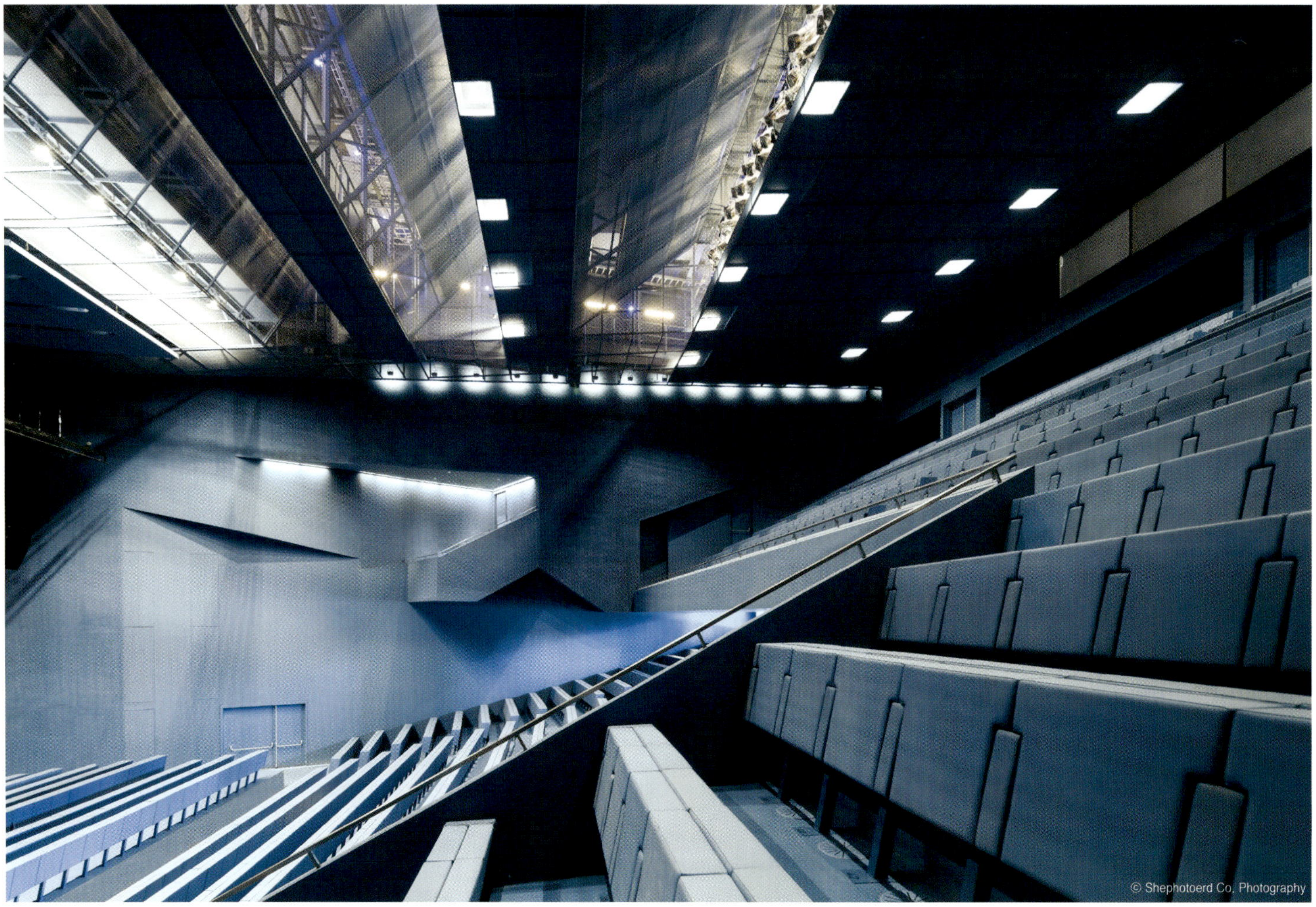

↑ Auditortum ↓ Auditortum

⌐¬ Escalator   ↑ Interior egress stair   ⌐¬ Corrugated glass

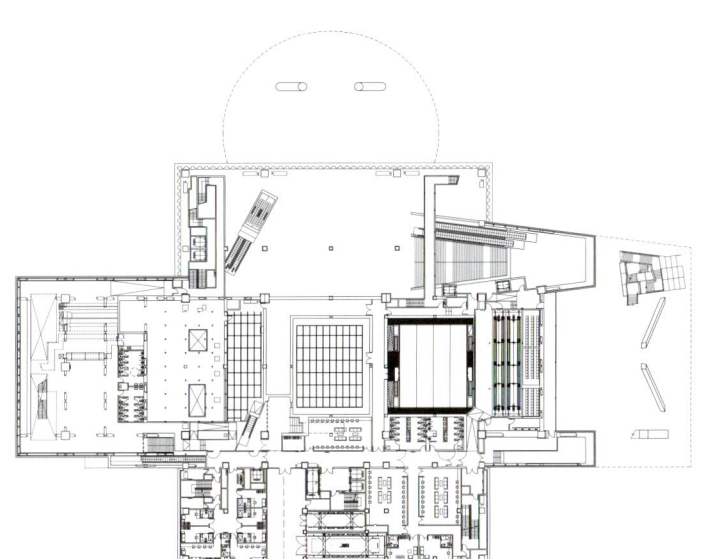

3RD FLOOR PLAN

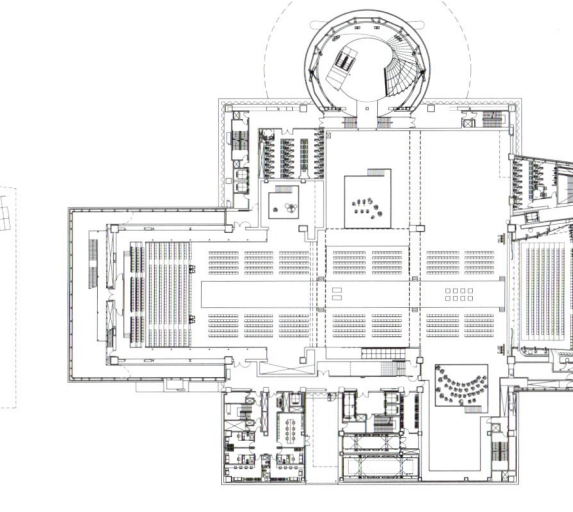

4TH FLOOR PLAN

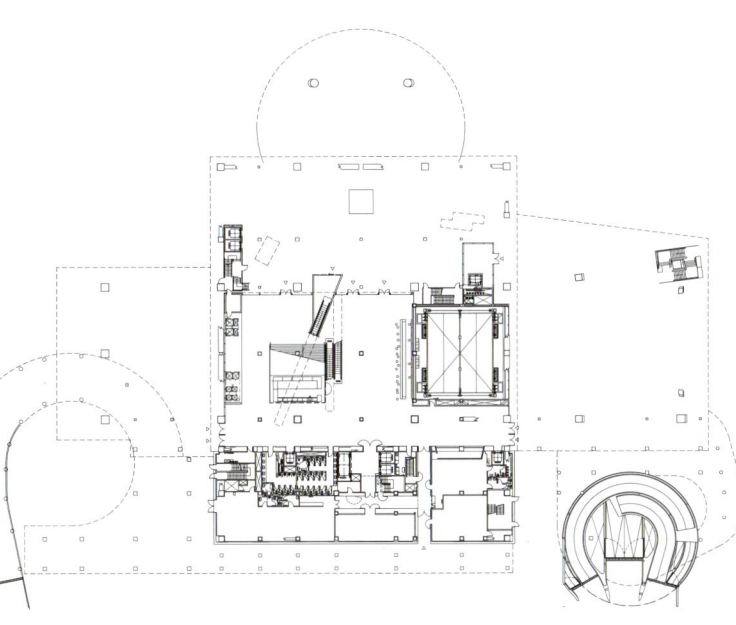

1ST FLOOR PLAN

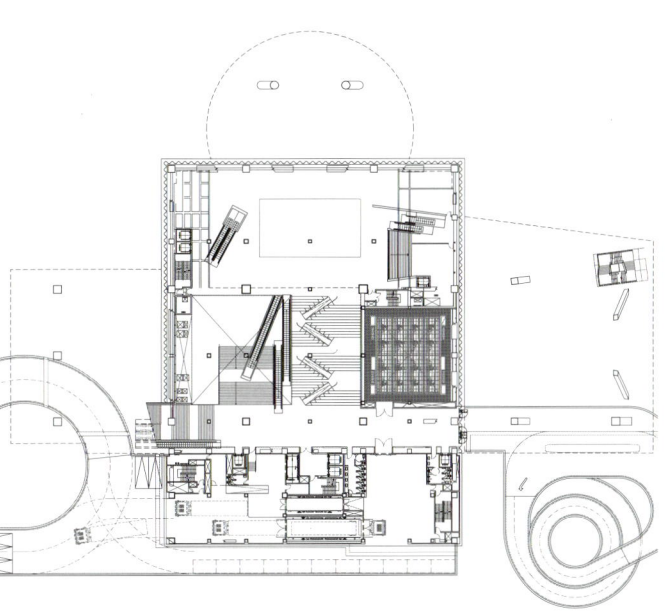

2ND FLOOR PLAN

↑ Auditorium

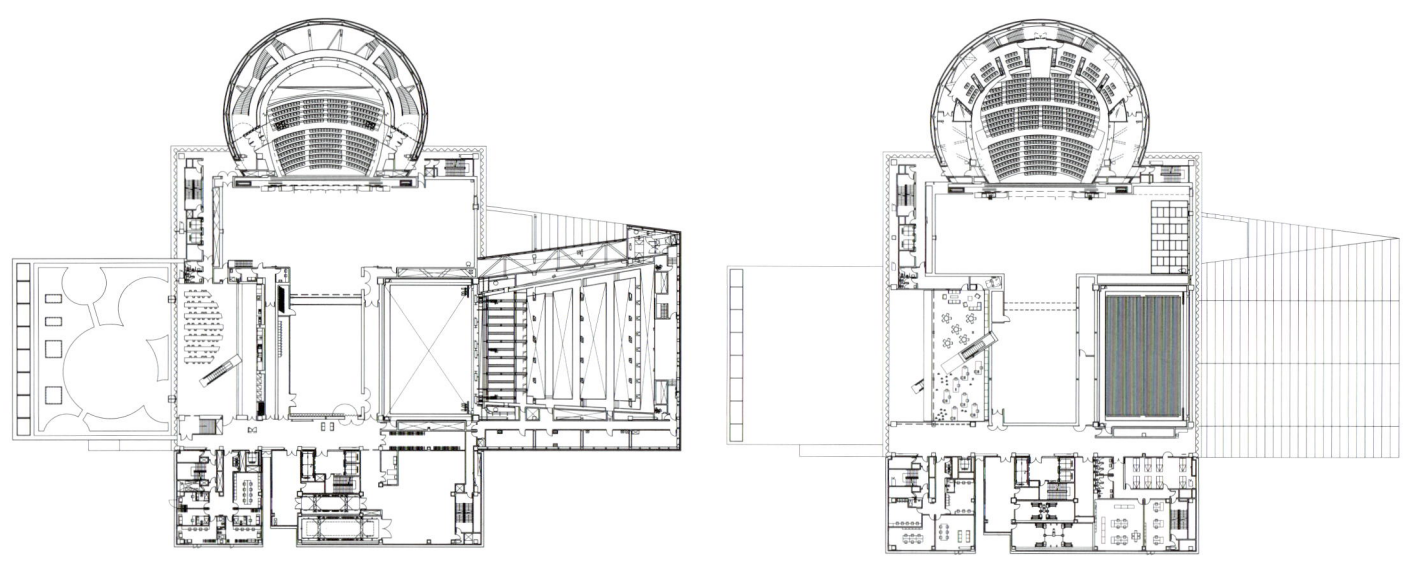

7TH FLOOR PLAN

8TH FLOOR PLAN

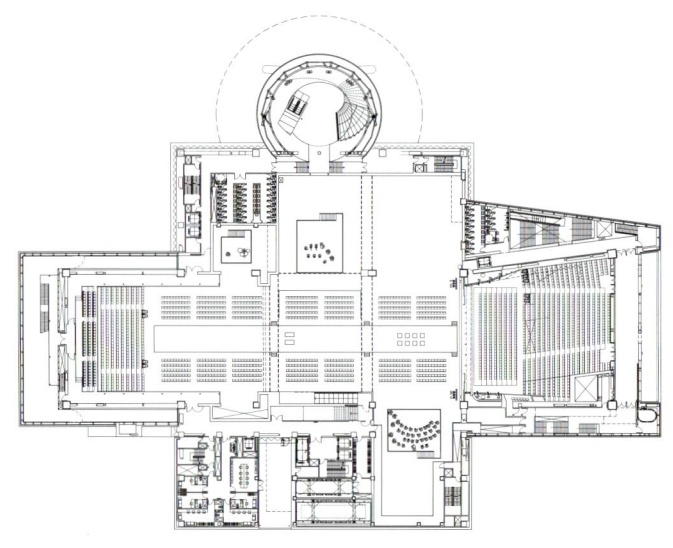
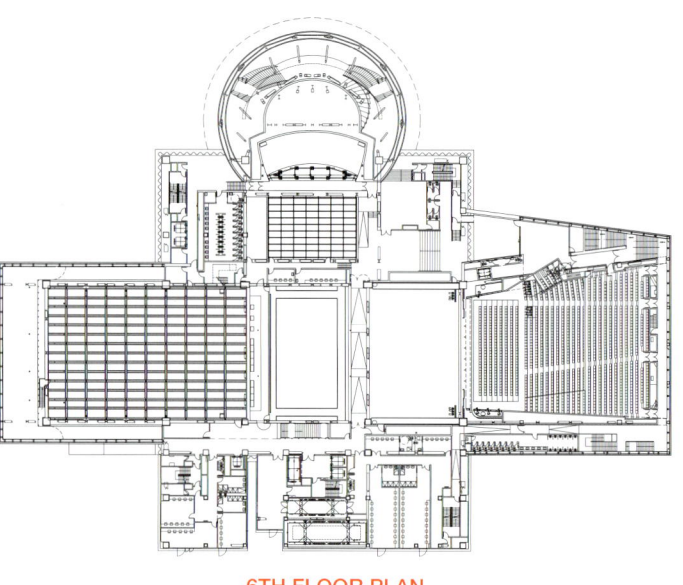

5TH FLOOR PLAN

6TH FLOOR PLAN

↑ Auditorium

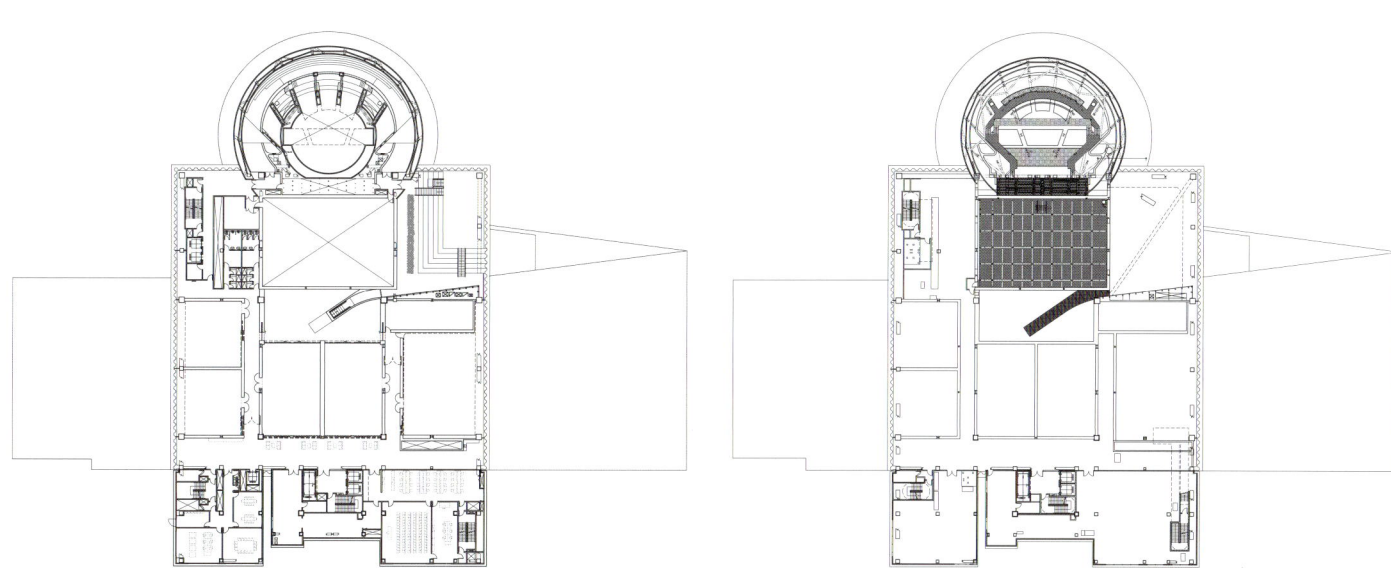

11TH FLOOR PLAN

12TH FLOOR PLAN

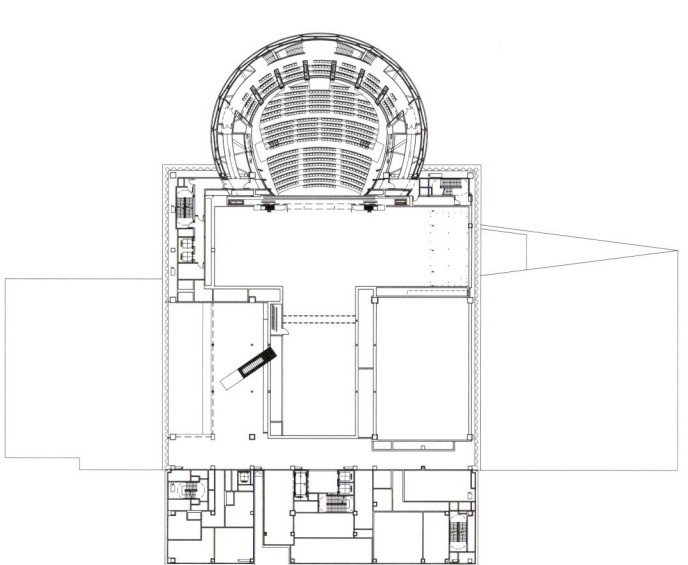

9TH FLOOR PLAN

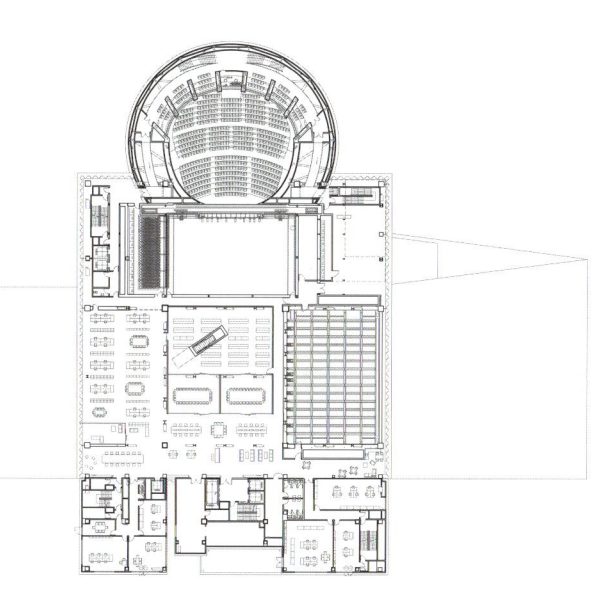

10TH FLOOR PLAN

높은계단집
STEPPING HOUSE

ARCHITECT : YOAP ARCHITECTS LTD. / INKEUN RYOO, DORAN KIM, SANGKYONG JEONG

HYANGDONG-DONG IS THE FIRST NEW TOWN IN GOYANG CITY in contact with the DMC area of Seoul, Susaek. Bongsan Mountain is a low mountain with an altitude of 200m, forming the border with Seoul. The block with the "Stepping House" lot is a block designed to create a riverside street in contact with the Hyangdongcheon Stream promenade. As it faces the streamside park, the biggest feature is that there are no restrictions on height, free from the restrictions of the due south sunlight line.

"Stepping House" is the tallest in the neighborhood. In particular, the stairwell and through-type elevator are connected to the rooftop, so that the rooftop can be used from the first floor and the underground shopping mall, so the stairway protrudes a lot from the outside and occupies the highest position like a signpost on streamside road in the Hyangdong district. In urban planning, the lots along streamside are induced to develop architecture by attaching both sides like cafe streets in Europe. Thanks to this, it has a height, but only the front and rear surfaces are exposed, so it seeks proportion with a 2D elevation design rather than a 3D design.

The Hyangdongcheon Stream runs through the front of the building, and it confronts the high-rise apartments in the Hyangdong district. Aside from judging whether it is right for urban planning to induce such an extreme difference in volume between multi-family and apartment areas, what we need to do is create our own beautiful skyline. When it comes to the word "own", the answer must be found somewhere in the middle between being unique and being reasonable. Columns and columns that are naturally created at the entrance of the first floor and basement are shaped as stairs, and they are applied to the skyline as they are. The module matched the one in the attic.

The underground promenade is connected at a natural level. However, you cannot enter through the front door of the house on the other side. Fortunately, a common external staircase was installed in consultation with the owner on the left, a common external staircase was installed so that it was possible to enter the house directly from the promenade. It does not overlap with the traffic line of vehicles, so you can easily use commercial and residential facilities while taking a walk safely.

In this project, a common entrance is to be planned at the rear of the building. The movement lines of an unspecified number of first-floor shopping malls and houses will overlap. How to solve the mixture of the piloti parking lot, the common entrance of the house, and the entrance to the shopping mall on the first floor? As an architectural solution, individual spaces such as landscaping, parking, and entrances, or public spaces and architectural elements (pilotis, common entrances, staircases, etc.) shall be faithfully designed, respectively. Even the area and height of houses and shops that can never be given up in commercial houses are made to exist in their best subsets rather than the overall combined proportion. Even the area and height of houses and shops that can never be given up in commercial houses exist in the best subset state rather than the total combined proportion. Believing in the proportion of each of the parts itself, elements should be separated and minimal design units must be created.

Location 79, Kkonnaeeum 2-gil, Deogyang-gu, Goyang-si, Gyeonggi-do, Republic of Korea **Use** Multiplex Housing **Site area** 232.10m² **Building area** 138.18m² **Gross floor area** 544.73m² **Completion** 2021 **Design team** Inkeun Ryoo, Doran Kim, Sangkyong Jeong, Chaelyul Kim **Contractor** First E&C Ltd. **Structural engineer** HANGIL structural engineering **Mechanical engineer** GM EMC **Telecommunication equipment** GM engineering **Photographer** Inkeun Ryoo

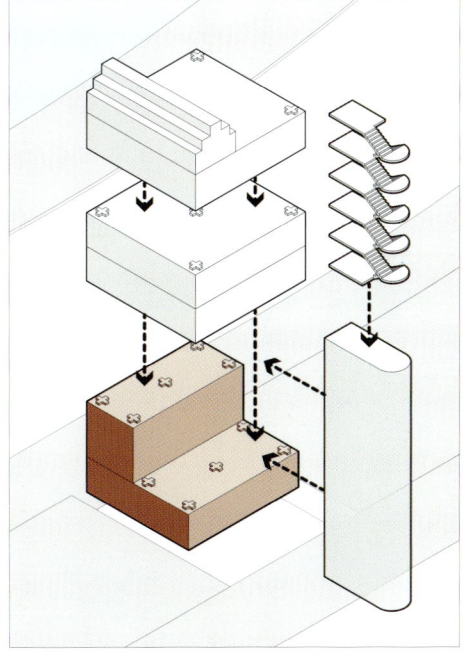

DIAGRAM

← Exterior view

↑ Street view

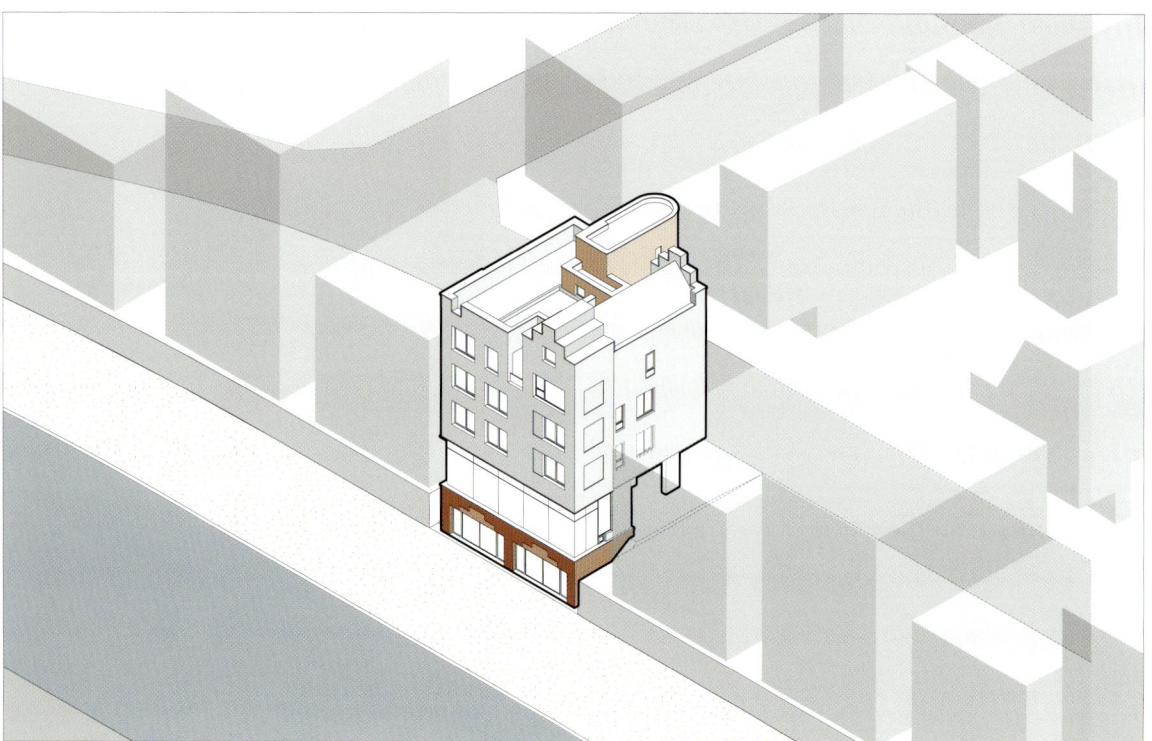

DIAGRAM

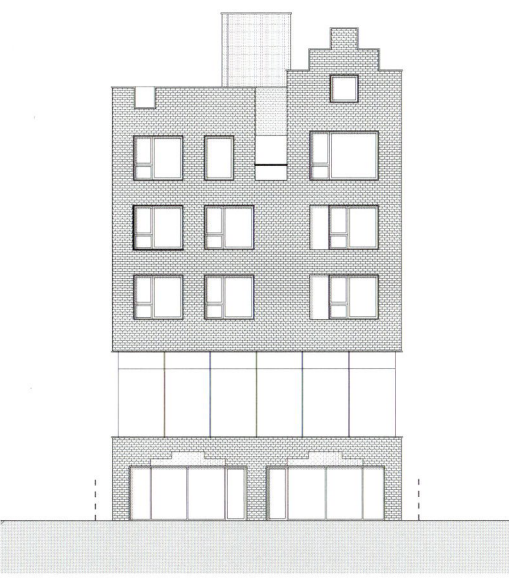

FRONT ELEVATION

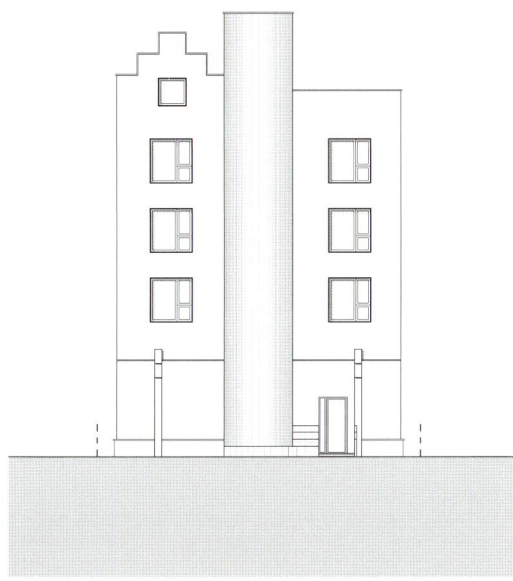

REAR ELEVATION

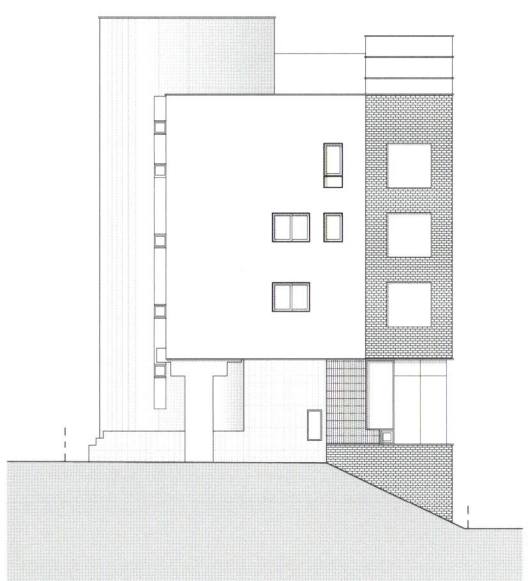

LEFT ELEVATION

↑ Partly view of exterior　↓ Partly view of exterior

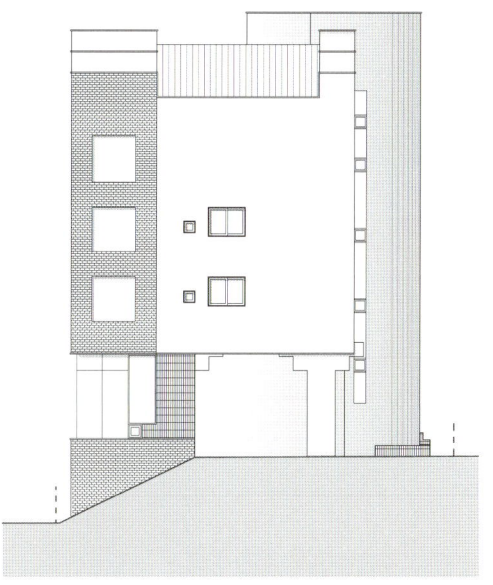

RIGHT ELEVATION

↑ Front view ↙ Main entrance ↳ Signage design

↑ Piloti

향동동은 서울의 DMC 지역, 수색과 접하는 첫 번째 고양시의 신도시이다. 봉산이라는 해발 200m의 낮은 산자락으로 서울과의 경계를 이루고 있다. '높은계단집' 필지가 있는 블록은 향동천변 산책로와 접하며 천변 가로를 조성할 수 있게 만들어진 블록이다. 천변공원과 면하다보니 정남일조 사사선의 제약에서 벗어나 높이의 제약이 없는 것이 가장 큰 특징이었다. '높은계단집'은 동네에서 높이가 가장 높다. 특히나 이 높이는 계단실과 관통형 엘리베이터가 옥상까지 연결되어 1층과 지하의 상가에서도 옥상을 이용할 수 있도록 했기 때문에 계단실이 외부에서도 많이 돌출되어 마치 향동지구 천변로의 이정표처럼 가장 높은 위치를 차지하고 있다. 도시계획에서 천변로의 필지들은 유럽의 카페거리처럼 양옆을 붙여서 건축을 개발하도록 유도되어 있다. 덕분에 높이를 갖지만 정면과 배면만이 노출되어 3D 설계보다 2D의 입면설계로 비례를 찾아간다.

건축물은 정면으로 향동천이 흐르고 향동지구의 고층 아파트와 대치된다. 도시계획이 다가구 지역과 아파트 지역이 이렇게 극단적인 볼륨의 차이가 나게 유도하는 것이 맞는지 판단하는 것은 차치하더라도 우리가 해야할 일은 나름의 아름다운 스카이라인을 만들어 주는 것이다. 나름이라는 말은 독특함과 합리적인 것, 그 중간 어디에서 답을 찾아야 한다. 1층과 지하의 진입에서 자연스럽게 만들어지는 단과 단들이 계단으로 형상화되고 그것들 그대로 스카이라인에 적용했다. 그 모듈은 다락의 모듈과 맞아떨어졌.

지하는 산책로가 자연스러운 레벨로 연결되어있다. 하지만 반대편 주택의 현관으로는 진입할 수 없다. 다행이 왼편의 건축주와의 협의로 공동 외부계단이 설치되어 산책로에서도 주택으로 바로 진입할 수 있게 설치되었다. 차량의 동선과 겹치지 않으니 안전하게 산책을 하다 상업시설과 주택의 시설을 쉽게 이용할 수 있다.

본 프로젝트는 공동현관을 건축물 후면에 계획해야 하는 경우이다. 불특정 다수의 1층 상가 동선과 주택의 동선이 겹치게 된다. 필로티 주차장, 주택의 공동현관, 1층 상가입구등이 혼재되는 것을 어떻게 풀어갈 것인가. 건축적인 해결책으로 조경, 주차, 현관 등의 개별공간 또는, 공공의 공간과 건축요소(필로티, 공동현관, 계단실 등)를 각각 충실하게 디자인 한다. 상가주택에서 절대 포기하지 못하는 주택과 상가의 면적과 높이까지도 전체 조합된 비례보다는 각각이 최상의 부분집합 상태로 존재하게 한다. 부분의 합 각각 자체의 비례를 믿고 요소를 분리하고 최소의 디자인 유닛을 만든다.

⌐┐ Piloti
← Entrance
⌐ Neighborhood facility

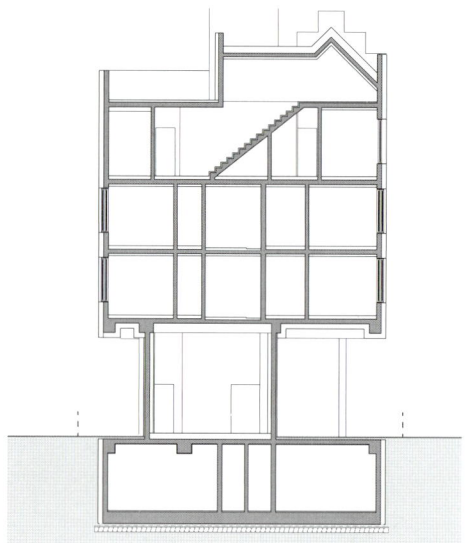
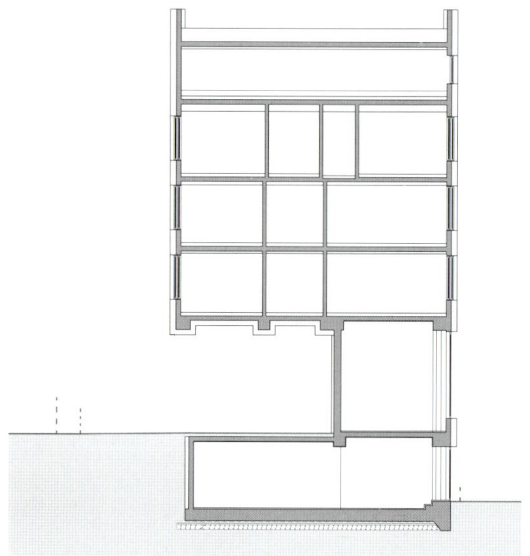
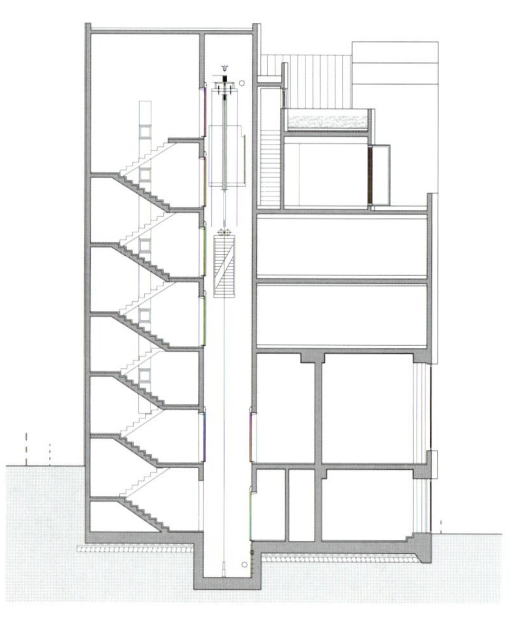

SECTION

↑ Neighborhood facility   ↓ Staircase

↖ Interior view   ↙ Living room   → Living room

1 NEIGHBORHOOD FACILITY
2 STORAGE
3 PARKING LOT
4 ELEVATOR HALL
5 BEDROOM
6 SUB KITCHEN
7 BATHROOM
8 KITCHEN
9 LIVING ROOM
10 DRESS ROOM
11 ATTIC
12 ROOFTOP TERRACE

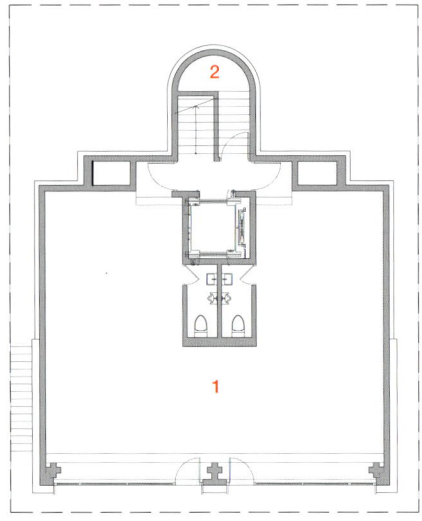

BASEMENT FLOOR PLAN

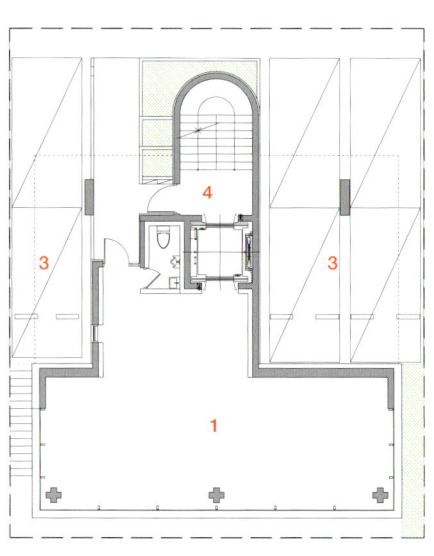

1ST FLOOR PLAN

← Stairs → Attic

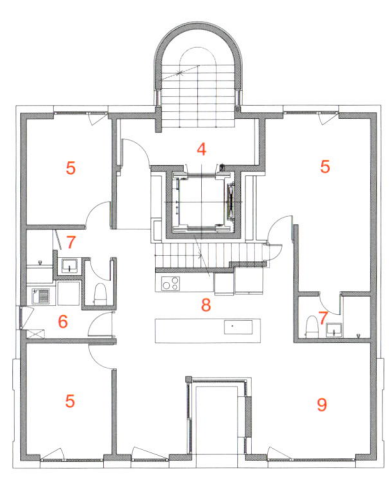

4TH FLOOR PLAN

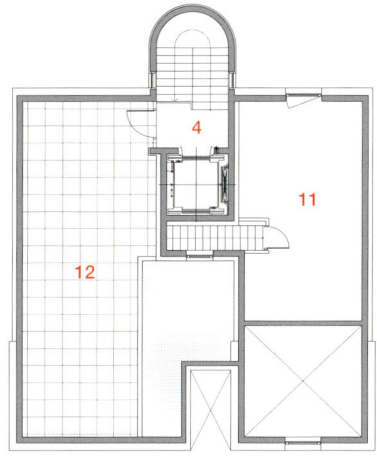

ATTIC FLOOR PLAN

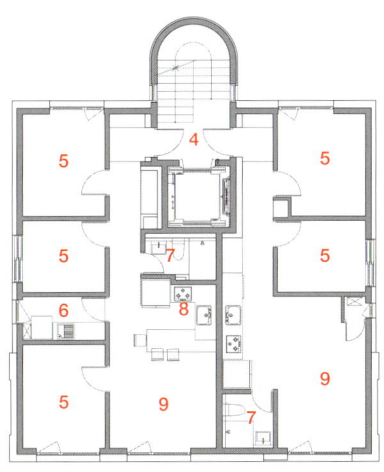

2ND FLOOR PLAN

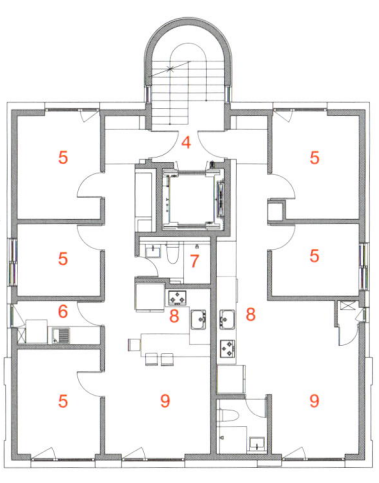

3RD FLOOR PLAN

얼리 비케이케이
EARLY BKK CAFE

ARCHITECT : SPACECRAFT CO., LTD. / NOPPACHAI AKAYAPISUD, SATHIKA JIENJAROONSRI

EARLY BKK STARTED WITH A STRONG INTENSION AND PASSION of Kaytita Chaisuksiri, to create a community cafe with truly green and sustainable concept in her neighborhood. She found a site location in her village where is a strong living community in Bangkok. In this community, mostly big families with all generations including pets, therefore Early BKK aimed to be a tangible space where can be accessible for all.

Unlike most of the cafe and retail business which always think about making profits. Early BKK, in the other hand, think about how to introduce recycle and green concept and lifestyle to the visitors and neighbors. Seeing many problems arising with regards to waste management these days, upcycling materials that have been overlooked are taken to important parts of the design.

Since statement of Early BKK is very clear, Spacecraft started to do the research about upcycling materials and trash around neighborhood. And found that the most familiar trash that can be recycle and reuse are different types of containers, such as milk cartons and glass bottles. The architects decided to use these waste containers as the main materials throughout the design, from architectural facade to interior finishing and furniture.

They worked together with a factory to produce a 're-board' which is a board made from 100 percent of small cut pieces of used milk cartons. Re-board was used in most in the interior space such as door, ceiling, chairs and tables. To control over all color scheme, we carefully selected color of used milk cartons only in orange and warm colors scheme, to match with handmade brick tiles floor.

Beer bottles are another main reused material we used in many different ways in this project. Obviously on the facade, around 600 used beer bottles were placed in the metal rings structure of the building. They created stunning effect of lights and shadow during days into an interior space, as well as tell good story of Early BKK's objective from the exterior appearance. Some of used beer bottles are broke down into pieces, and used them as main component of terrazzo counter working top and toilet floor. Some walls are made by bumping bottles into concrete wall, creating interesting wall patterns and texture. We called this "Bottles fossil" since the production and final look is similar to the fossil by leaving traces on the wall. Moreover, we added some small details by using trash such as a used galvanized iron oil tank as a basin counter , and used glass bottles as a door knob.

Early BKK is a small 2-storey building in the corner of road. The entrance at the corner is a small double space courtyard with a big tree and pet parking area, offering a welcoming and friendly vibe. 1st floor is a main coffee bar including coffee machines and slow bar, with few seats. Retails shelves for refill products, home coffee equipment, pet products and some second-hand clothes are in the back of coffee bar area. 2nd floor is all seating area, designing with large window on both sides for well ventilation.

In the end, Early BKK are an inspiring showcase of how interesting upcycling and waste materials can finally created. People might come here for a coffee, but will definitely go back with some changes of their mindset about waste and recycle concept. Believing in small intension can make the world a better place, Early BKK prospected to be a small part of creating some changes in the society.

Location Bangkok Thailand **Use** Cafe **Site area** 120m² **Completion** 2022 **Structural Engineer** Aphichart Wongdee **Photographer** Thanapol Jongsiripipat

← Facade view ↱ Recycle bottle ↳ Bottle fossil

↑ Exterior view

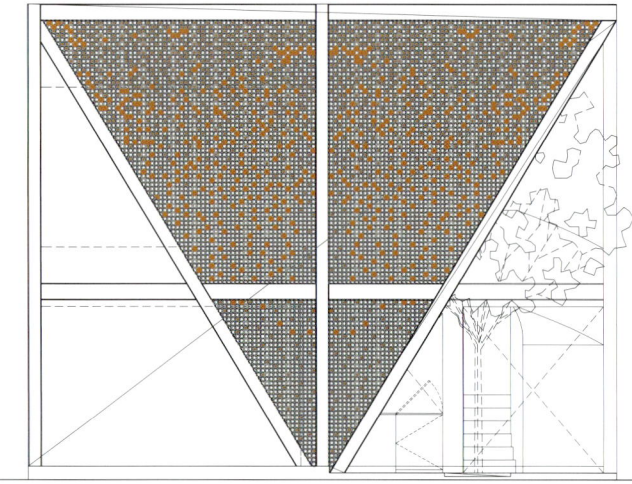

FRONT ELEVATION

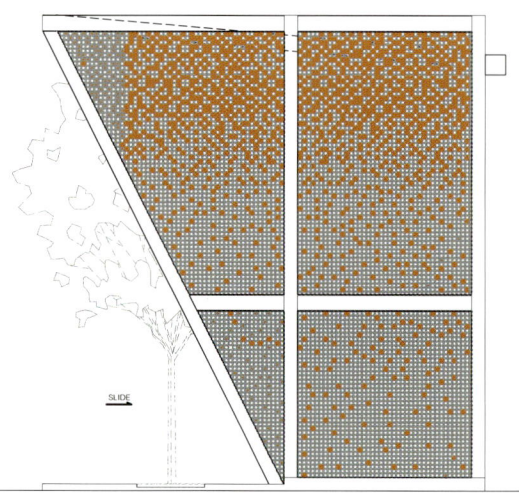

RIGHT ELEVATION

↑ Street view

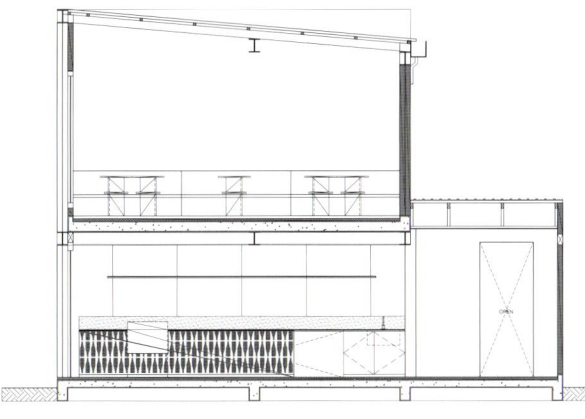

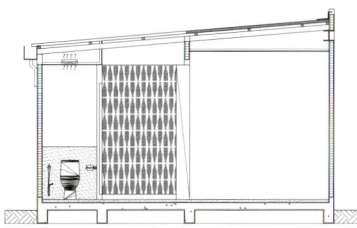

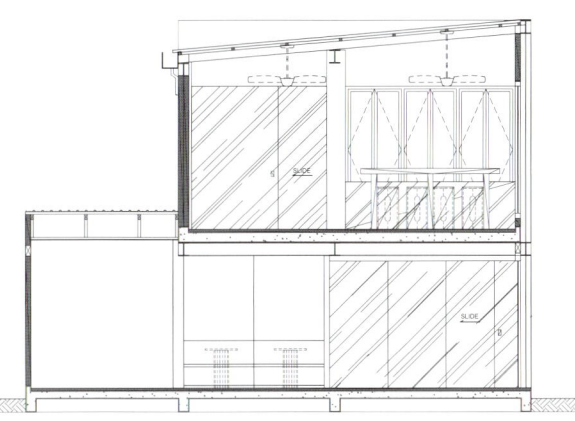

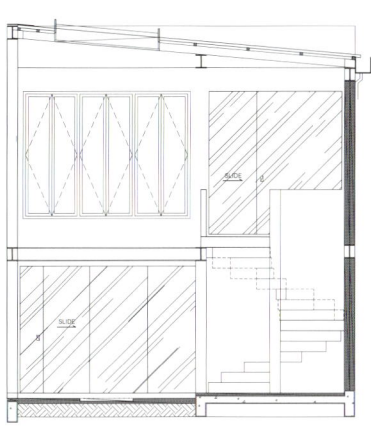

SECTION

↑ Hall on the 2nd floor   ↓ Hall on the 2nd floor

얼리 비케이케이는 카이티타 차이수크시리의 강한 의지와 열정으로 그녀의 동네에 친환경적이고 지속 가능한 커뮤니티 카페를 만들기 시작했다. 그녀는 끈끈한 생활 공동체를 이루는 방콕의 마을을 부지로 삼았다. 그 마을은 반려동물과 전 세대가 소속된 대가족이 대부분이라 얼리 BKK는 모두에게 열린 유형의 공간을 조성하고자 했다.

이익 실현만을 염두에 둔 대부분의 카페나 소매상점과 달리 얼리 비케이케이는 방문객과 이웃들에게 재활용의 가치와 친환경적인 라이프스타일을 선보이고자 했다. 폐기물과 관련해 많은 문제가 발생하는 현실을 반영해 그간 간과되어 온 업사이클링 개념을 디자인의 테마로 채택하였다.

얼리 비케이케이의 명확한 비전에 따라 스페이스크래프트는 동네의 쓰레기 처리 및 업사이클링 현황을 파악하기 시작했다. 그러면서 재활용 및 재사용이 가장 활발하게 이루어지는 쓰레기가 우유 팩 및 유리병과 같은 다양한 형태의 용기라는 사실을 발견했다. 따라서 파사드부터 인테리어, 가구에 이르기까지 설계 전반에 걸쳐 이러한 폐기물 용기를 주요 소재로 활용하였다.

폐기물 우유 팩 조각 100%로 이루어진 보드 즉, 리보드(re-board)를 생산하는 공장과 협력해 문, 천장, 의자 및 테이블 등 실내 인테리어 대부분에 다양하게 접목했다. 색상의 톤을 맞추기 위해 수제 벽돌 타일 바닥과 조화를 이루는 주황색 및 따뜻한 색감의 우유 팩으로 이루어진 보드를 신중하게 선택했다.

이 프로젝트에서 다양한 부분에 주로 사용된 재사용 재료로 맥주병이 있다. 고리로 연결된 메탈 구조물에 약 600개의 재활용 맥주병을 조합해 파사드에 장식했다. 낮에는 빛과 그림자의 대비로 내부 공간에 놀라운 효과를 연출하며, 외부로는 얼리 비케이케이의 지향점을 상징하는 역할을 한다. 재활용 맥주병 일부는 잘게 부수어 테라조 작업대와 화장실 바닥의 주요 요소로 활용했다. 콘크리트 벽 일부에는 병을 부딪쳐 독특한 패턴과 질감을 연출했다. 벽면에 흔적을 남기는 과정과 최종 결과물이 화석과 비슷하게 보여 '병 화석'이라고 부른다. 또한 아연 도금된 철제 오일 탱크와 같은 폐기물을 세면대로 재사용했으며, 문손잡이에는 유리병을 장식해 디테일을 살렸다.

본 건축물은 도로 코너에 있는 작은 2층 건물이다. 코너 입구에는 큰 나무를 심고 반려동물을 위한 소규모의 이중 공간을 안뜰로 구성해 방문객에게 환대와 친근한 분위기를 제공한다. 1층에는 커피 머신과 슬로우 바로 메인 커피 바를 구성하고, 적은 수의 의자와 테이블을 놓았다. 커피 바 뒤쪽 선반에 판매를 위한 리필제품, 가정용 커피 머신, 반려동물 제품 및 일부 중고 의류를 배치했습니다. 2층은 좌식 공간이며, 통풍이 잘 되도록 양쪽에 큰 창을 설계하였다.

결국 얼리 비케이케이는 업사이클링과 재활용 소재를 활용해 굉장한 것을 만들어 내는 것으로 영감을 주는 쇼케이스다. 사람들은 이곳에 커피를 마시러 올 수도 있지만, 돌아갈 때는 폐기물 및 재활용에 대한 생각의 변화를 가지게 될 것이다. 작은 의지가 세상을 더 나은 곳으로 만들 수 있는 믿음으로 얼리 비케이케이는 사회의 작은 변화를 만들어 갈 것을 기대한다.

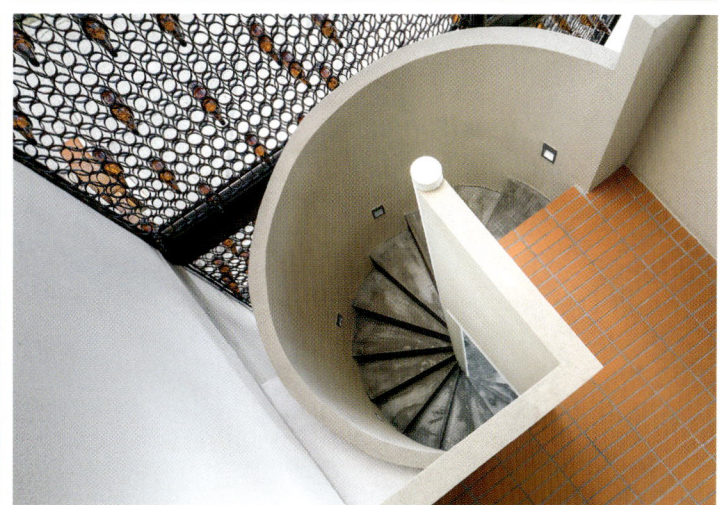

← Coffee station → Seating area ↙ Top view ↘ Spiral stair

1 CAFE AREA   3 SPIRAL STAIR   5 RESTROOM   7 FOYER
2 KIOSK   4 RETAIL CORNER   6 SEATING & WORKSHOP AREA

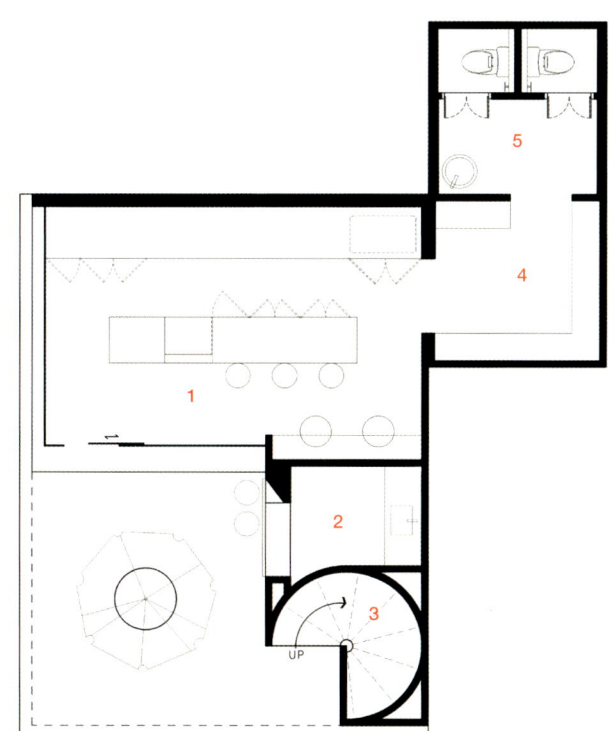
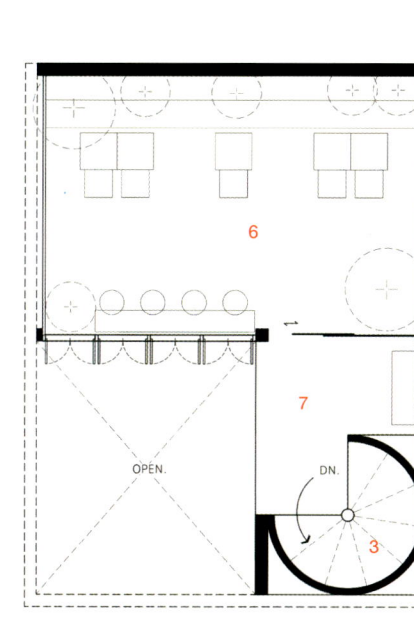

1ST FLOOR PLAN **2ND FLOOR PLAN**

스트리트 앤 가든 아파트

STREET AND GARDEN APARTMENTS

ARCHITECT : rh+ architecture / ALIX HÉAUME, ADRIEN ROBAIN

THE PROJECT IS LOCATED IN THE 11TH DISTRICT OF PARIS. The building engages in a delicate dialogue with its built surroundings, with the larger building of 45 meters long for 9 meters wide. The main white facade is stepped so that it follows on from the neighboring buildings, one of which is set further back. On the garden side, the building is clad in solid wood modules that give it a warm, intimate, peaceful atmosphere that contrasts with the hustle and bustle of the Paris faubourg. The facades are composed of blocks Concrete facade 20x20cm wood, solid larch 3, 5 and 7cm thick. These blocks are wedged to create a game of random masses and bring shadows, vibrate the facades. On the street side, the facade is made of white concrete, complexion in the mass: white is not everything is uniform and has traces specific to "poured in place" concrete. some parts are "hammered", a rough texture that highlights openings and the inclined facade, which is a link with the neighboring building set back.

All the apartments are floor-through so that they make the most of the calm green space set back from the street. The heights of the buildings have been deliberately calculated to avoid impressions of density and to ensure the apartments, including those at garden level, receive as much sunlight as possible. This project delicately responds to the question raised by the occupation of a deep, narrow plot in the very heart of a busy area.

The construction of two buildings separated by gardens was the best performer in terms of ventilation, land use and views. The majority of dwellings have two opposite facades and a balcony or a terrace. The porch and the open passage allow to guess the garden from the street. The density of these two small buildings of 14 dwellings, located between two gardens, is reasonable and balanced: 9 units on the street side, with terraces the 3rd floor, and 5 units on the garden side, with terraces or balconies on all floors. Roofs are broken to create terraces and let in the light in the nearby courtyards. The courtyard of the building is quiet and airy, with a very present vegetation.

Location Paris 10e, France **Use** Housing **Site area** 475 m² **Built area** 307m² **Gross Floor area** 1,100m² **Project Manager** Alix Héaume **Design team** rh+ architecture **Contractor** Brezillon **Photographer** Luc Boegly

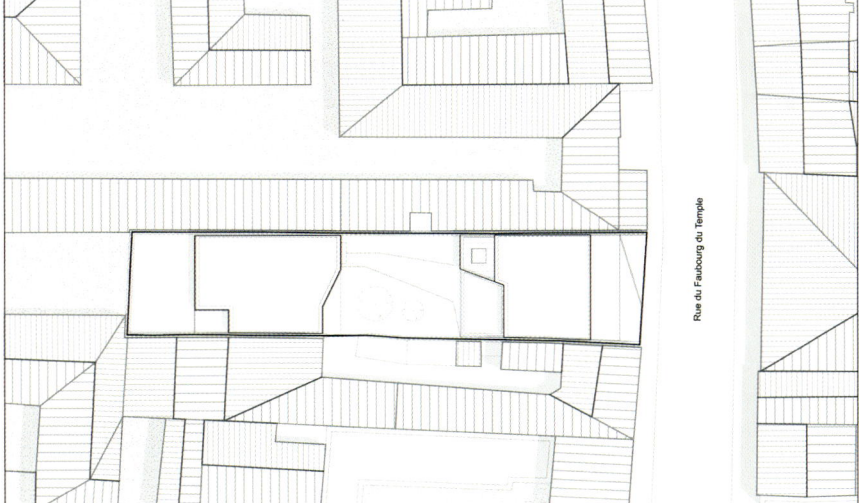

SITE PLAN

← Facade view → Exterior & street view

On the garden side facade

이 건물은 파리 11구에 위치해 있다. 길이가 45m 더 길고, 폭이 9m 더 넓은 큰 빌딩이 있는 주변 환경과 자연스럽게 어우러져 있다. 화이트 색상의 메인 파사드는 계단식으로 되어 있어 주변 건물들과 이어지고, 건물 중 하나는 더 뒤쪽에 있다.

정원 쪽 건물은 단단한 목재로 마감해 파리 포부르의 번잡한 분위기와 대비되는 평화롭고 따뜻하며 친근한 느낌을 준다. 파사드는 블록과 콘크리트 파사드, 20x20cm의 목재와 3, 5, 7cm 두께의 낙엽송으로 구성되어 있다. 이 블록들은 다양한 질감과 그림자 효과를 연출하며 한층 생동감 있는 파사드를 구현한다. 거리 쪽 파사드는 균일하지 않은 화이트 톤으로 이루어져 마치 콘크리트를 "때려 부은 듯한" 느낌을 준다. 일부에는 "망치질"로 거친 질감을 더해 문이 열리는 부분을 강조하고, 살짝 경사진 느낌을 주어 뒤쪽에 있는 주변 건물로 이어진다.

하나의 층 전체를 아파트로 구성해 분주한 거리에서 떨어진 고요한 녹지 공간을 최대한 활용할 수 있게 되었다. 건물의 높이는 번잡하지 않은 인상을 주고 정원 층을 포함한 아파트에 최대한 빛이 많이 들어올 수 있도록 계산되었다. 이 건물은 번잡한 지역 한가운데 깊고 좁은 부지를 점유하는 데서 발생하는 문제를 해결해야 했다. 그래서 두 개의 건물을 정원으로 분리해 부지 활용도와 환기 및 전망을 향상했다. 개별 가구 대부분 양쪽에 파사드가 있고 발코니와 테라스가 각각 하나씩 있다. 거리에서 현관과 열린 통로를 통해 가든의 위치를 알 수 있다. 두 개의 정원 사이에 세워진 두 건물에 14가구가 균형 있게 위치한다. 거리 쪽에 있는 9가구 중 3층에는 테라스가 들어가고, 정원 쪽에 있는 5가구는 전 층에 테라스 혹은 발코니가 있다. 지붕을 분리해 테라스를 조성했으며, 동시에 근처 안뜰로 빛이 들어온다. 빌딩의 안뜰은 조용하고 환기가 잘 되며 식물이 무성히 자라고 있다.

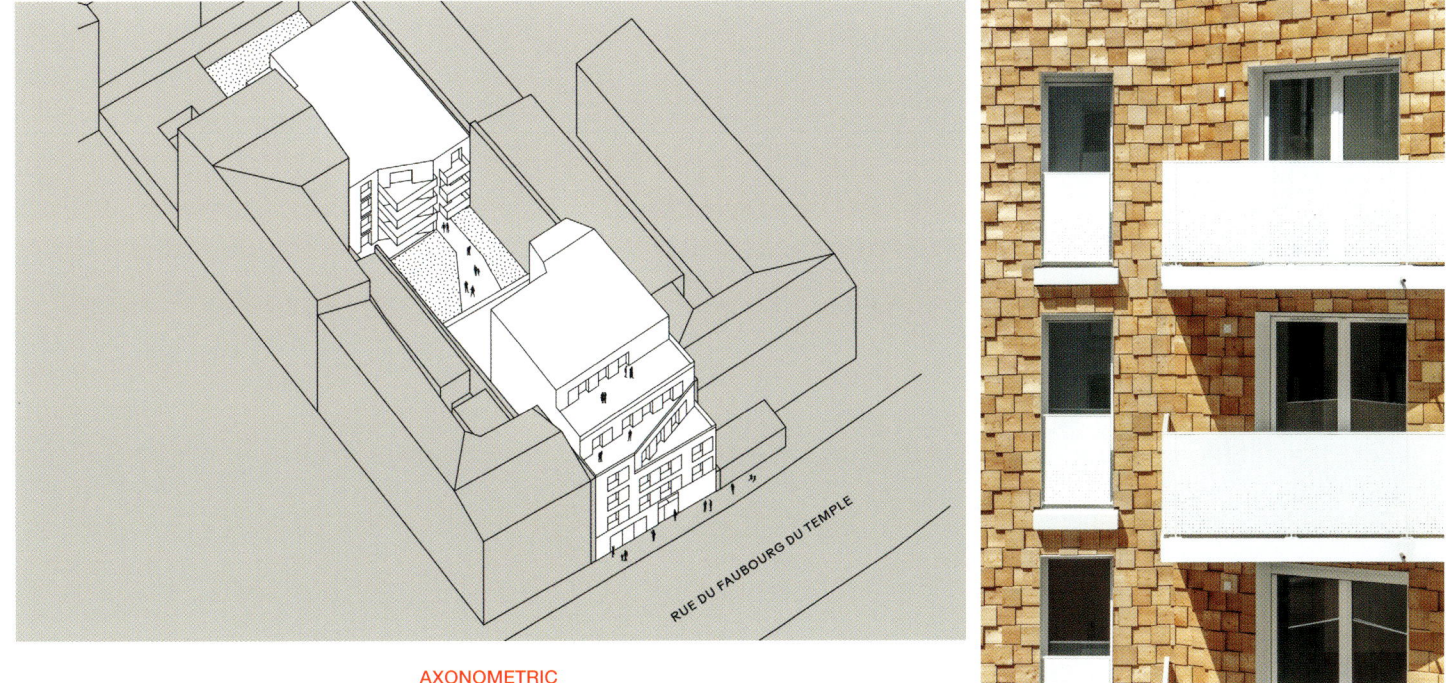

← Two buildings separated by gardens → On the garden side is clad in solid wood modules ↳ Wood modules facade & balcony

AXONOMETRIC

↑ Balcony of top floor ↵ Window & door of balcony ↳ Wood modules finish

↖ Street side facade   ↑ Balcony   ↗ Wood modules & window

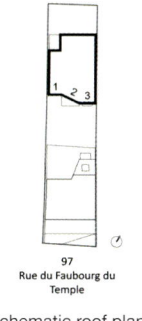

97
Rue du Faubourg du Temple

Schematic roof plan

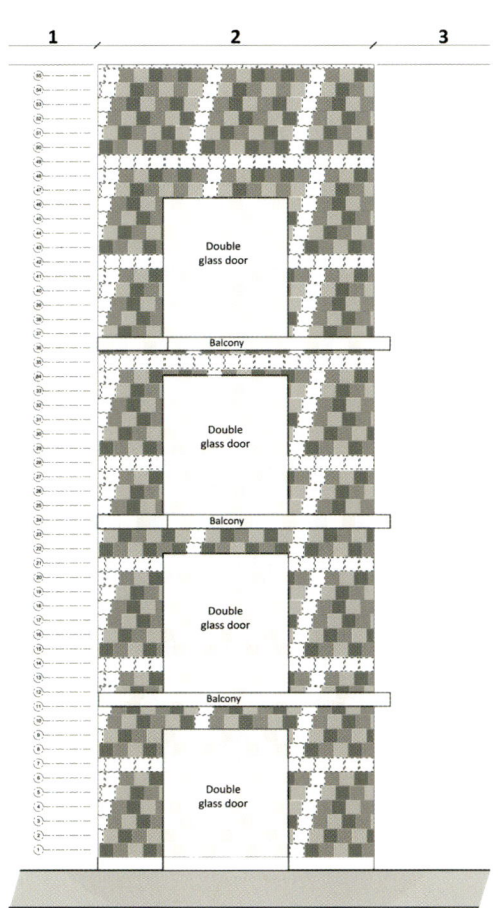

South elevation 1/20 : example of wood panels layout

Assembly of panels made of wooden blocks of three different depths identified as A, B, C.
One panel = 6 rows of 6 wooden blocks

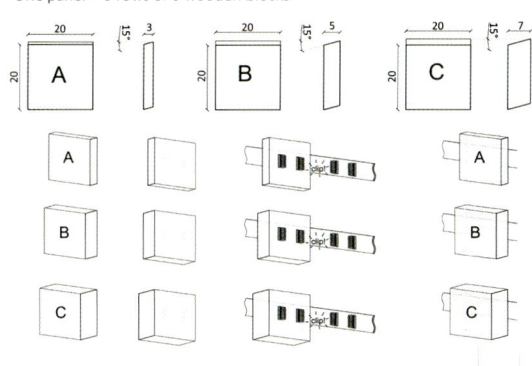

Types of wooden blocks and fixation

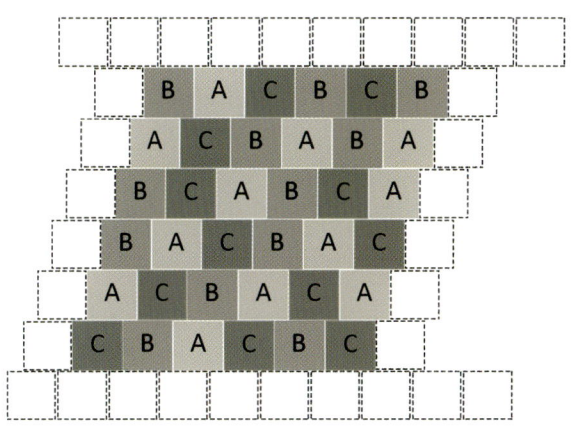

Detail of panel

DETAIL OF ELEVATION

↑ Interior space   ↵ The window system   ↓ Balcony & window   ↳ Colorful corridor

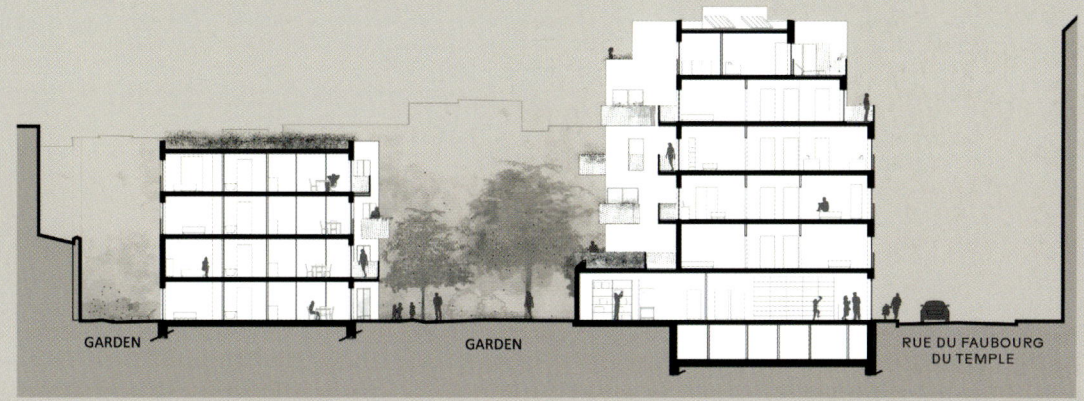
LONGITUDINAL SECTION A

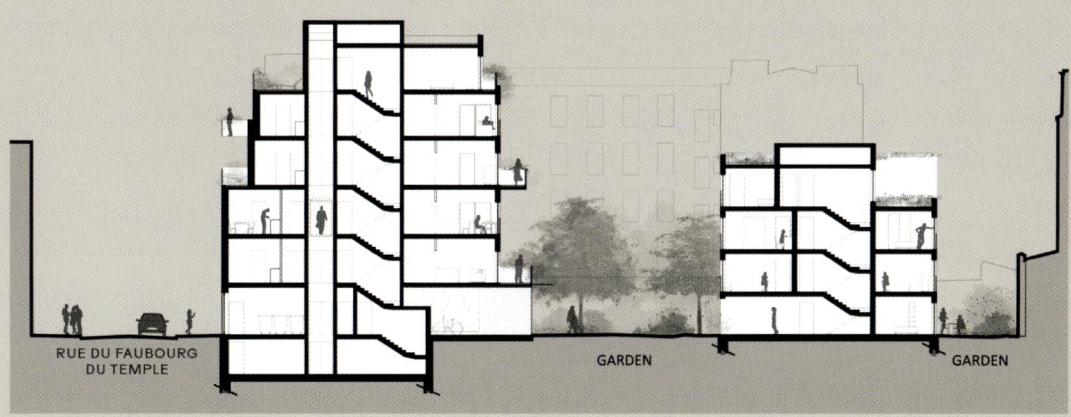
LONGITUDINAL SECTION B

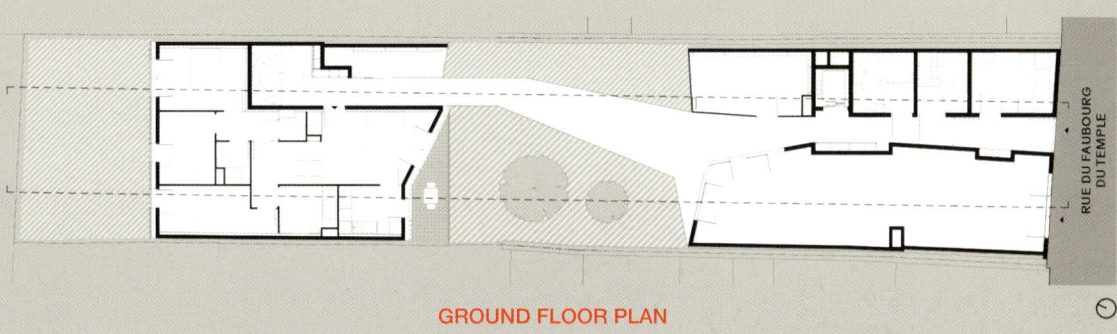
GROUND FLOOR PLAN

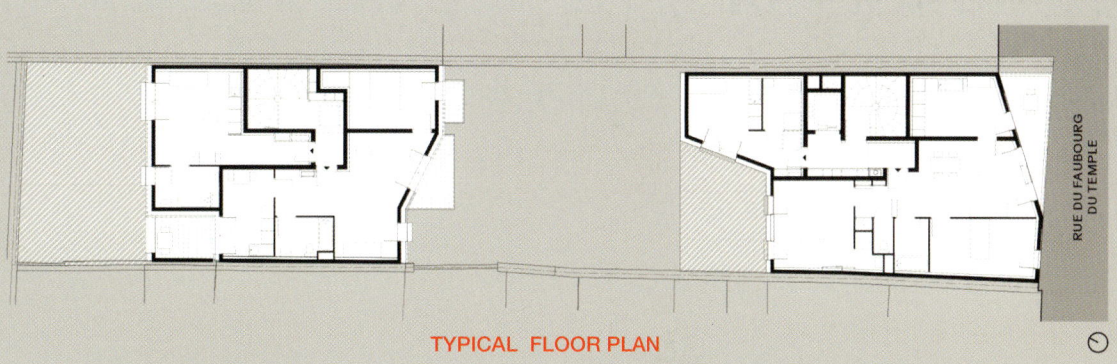
TYPICAL FLOOR PLAN

만화가의 집

A JAPANESE MANGA ARTIST'S HOUSE

ARCHITECT : TAN YAMANOUCHI & AWGL / TAN YAMANOUCHI

ON A SMALL NARROW PLOT (4.9M-WIDE, 14.7M-DEEP) in metropolitan Tokyo sits a wooden house. The clients are an up-and-coming manga artist, her partner, and two owls who are the new additions to the family. The manga artist made three requests: First, the house should accommodate the entire process of the artist's work, from creative concept to completion, meetings, and giving media interviews. Second, the house should be compact and should not open to the outdoors too much. Lastly and most importantly, the house should spark inspiration for creativity. Envisioned as "a building that floats a few centimeters above our daily lives," the architects strived to ensure that the dwelling is still tied to our tangible daily life but evokes a sense of fictional narrative.

The project began by selecting a plot closely linked to manga artists. The west facade facing the front road is designed as a warped seismic wooden wall evoking earth rising dynamically from the ground, with a tunnel penetrating the wall leading to the entrance. The opening in the solid wall invites visitors to step out from their ordinary and into the extraordinary, evoking a scene in a movie. The overall plan is designed to maximize the narrow and long plot of land, comprising the east area with three floor levels at the rear of the site and the west area with two floor levels to create a split-level floor plan. We rearranged the sequence of alternating split levels and created substantial differences in elevation. A light court (1.2m-wide, 5.5m maximum height) is provided on the north side of the house, and the number of other openings were limited as much as possible to create a contrast between light and dark in a "void" spreading throughout the house like an amoeba. A stairway runs up through the void space filled with contrasts of high/low and light/dark, making a three-dimensional composition that evokes a narrative experienced through the body.

The lifestyles of manga artists have changed dramatically during the COVID-19 pandemic. We no longer see assistants gathering at an artist's atelier since online production has become the mainstream of manga creation. Alternatively, their private houses have become a place for meetings and serve as studios for interviews and accommodating the press. The work of a contemporary manga artist could be divided into three stages: 1. Creation (enclosed and secluded), 2. Meetings (partially opened to others), and 3. Giving interviews (opening to the public), and the shades of public/private required in each stage varies. They designed a void with contrasts of high/low and light/dark to provide unenclosed yet defined areas, or what they call "ponds and banks," that would allow for subtle and flexible use of space in a very compact dwelling. We strive to continue creating architecture that embraces open narratives while pursuing logical solutions.

Location Tokyo, Japan **Use** Housing **Site area** 74.0m² **Building area** 44.16m² **Gross floor area** 86.45m² **Completion** 2022 **Graph Studio** Yuko Mihara **Contractor** Taishin Kensetsu / Yasuhiro Ikebe, Keisuke Nishide **Photographer** Katsumasa Tanaka

 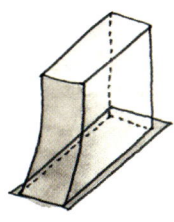 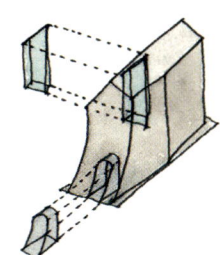 

CONCEPT DEVELOPMENT

← Front facade

Bird's eyes view

대도시 도쿄에 위치한 너비 4.9m, 폭 14.7m의 작고 좁은 부지에 세워진 목조 주택이다. 이곳에 주거하는 가족은 인기를 끌고 있는 만화가와 그녀의 파트너 그리고 두 마리의 올빼미다. 건축주의 세 가지 요청은 다음과 같다. 첫째, 이 집에서 창작물 구성부터 완성은 물론 미팅과 미디어 인터뷰에 이르는 다양한 작업이 가능해야 한다. 둘째, 집의 규모는 크지 않고 외부로부터 어느 정도 폐쇄성이 있어야 한다. 마지막으로 가장 중요한 것은 창조적인 공간에서 영감을 얻을 수 있어야 한다. 디자인의 테마는 일상과 밀접하게 연결되어 있으면서도 허구적인 내러티브를 부여하여 '일상 위에 구름처럼 떠 있는 집'으로 결정되었다.

공간을 설계할 때 만화가의 라이프스타일에 초점을 두었다. 전면 도로에 면한 서쪽의 파사드는 뒤틀린 목재로 내진벽을 구성해 땅에서 다이내믹하게 솟아오른 대지를 구현했으며, 진입로로 이어지는 벽을 관통하는 터널을 더했다. 단단한 벽이 열린 곳으로는 마치 영화의 한 장면처럼 방문객이 일상에서 벗어나 특별한 공간으로 초대받게 된다. 전체적으로 좁고 긴 대지를 최대한 활용할 수 있도록 설계했으며, 대지 후면 동쪽으로는 3개 층을, 서쪽으로는 2개 층을 세워 분할된 구조를 완성하였다. 교대로 분할되는 층으로 재구성해 높이가 균일하지 않은 실루엣이 연출된다. 집의 북쪽에는 너비 1.2m, 최대 높이 55m의 채광 공간을 구성하고 열려 있는 부분을 최소화하여 전체적으로 '비어 있는' 공간에 빛과 어둠이 대비를 이루며 아메바를 연상시키는 공간을 형성하였다. 계단을 통해 높이의 고저와 빛과 어둠의 대비가 뚜렷한 빈 공간을 오르내리며 육체적으로 경험하는 입체적인 내러티브가 심화된다.

코로나19 팬데믹으로 인해 만화가의 라이프스타일은 극적으로 변화했다. 이제는 만화 제작이 주로 온라인을 통해 이루어지고 있기 때문에 만화가의 작업실에 어시스턴트가 상주해야 할 필요가 크지 않다고 보았다. 대신 사적인 주택 공간에서 회의가 이루어지고 언론과 대면하는 스튜디오의 역할이 요구되었다. 현대 만화가의 작업은 1. 창조(밀폐 및 고립), 2. 회의(외부에 부분 개방), 3. 인터뷰(대중에 공개)의 세 단계로 진행되기 때문에 공개/비공개의 경계를 분명히 할 필요가 있다. 높이의 고저와 빛과 어둠의 대비를 강조한 공간은 폐쇄되어 있지 않은 특정 공간으로 '연못과 제방'의 역할을 하며 작은 주택 공간을 한층 효율적으로 활용할 수 있도록 합니다. 건축가는 논리적인 구조를 통해 개방적인 내러티브를 제공하는 건축을 지향했다.

← Exterior view   → Street view

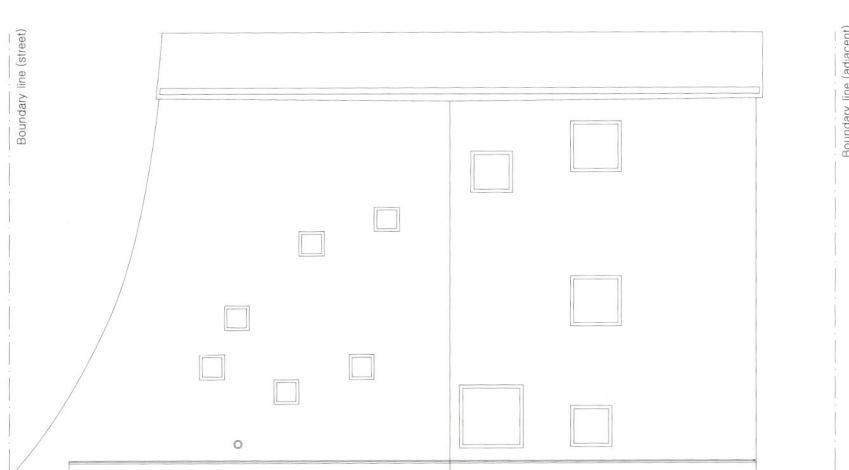

SOUTH ELEVATION

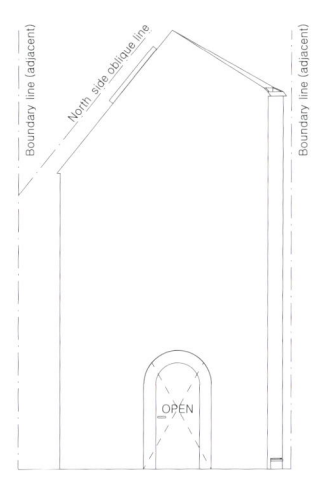

WEST ELEVATION

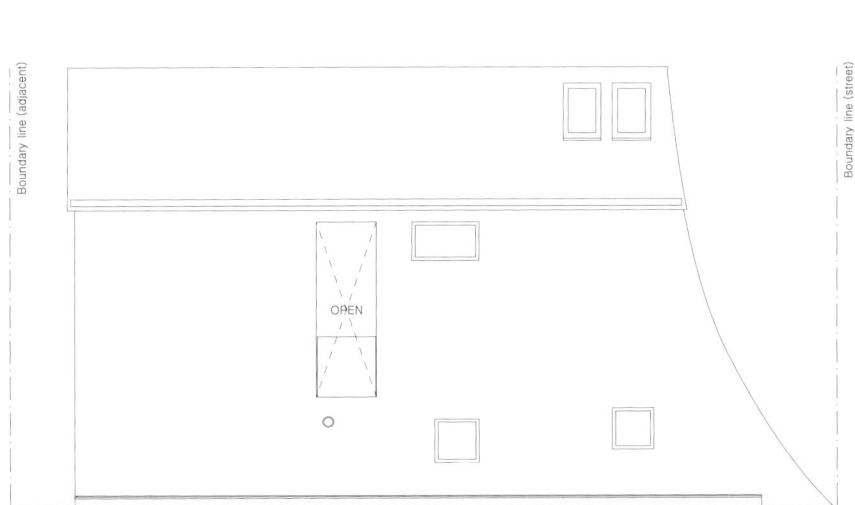

NORTH ELEVATION

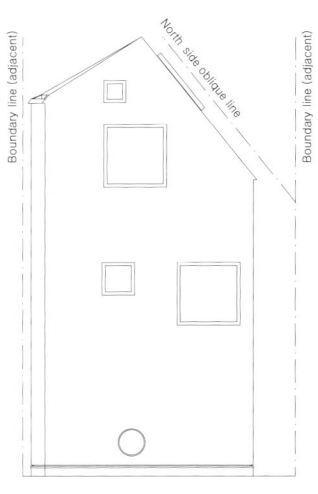

EAST ELEVATION

↑ Interior view   ↙ Stair   ↳ Stair

← Dining area   → Dining area

1. EEXTERIOR WALL : T30 EARTHEN WALL, T12 STRUCTURAL PLYWOOD, T18 VENTILATION LAYER, BREATHABLE WATERPROOF SHEET, T50 THERMAL INSULATION, T9 STRUCTURAL PLYWOOD, T105 GLASS WOOL, AIRTIGHT WATERPROOF SHEET
2. EP, T12.5 GB
3. OWL : WHITE-FACED SCOPS OWL
4. DIATOMACEOUS EARTH CEILING, T12.5 GB
5. OWL : INDIAN EAGLE-OWL
6. T1 BLACK STEEL PLATE, T12.5 WATERPROOF BOARD
7. EP, T12.5 GB, T105 GLASS WOOL
8. PLANT : BOTTLE TREE
9. VENTILATION GARARI
10. T15 OAK WOOD FLOORING (HERRINGBONE), T24 STRUCTURAL PLYWOOD, T300 GLASS WOOL
11. T15 IPE WOOD FLOORING, FRP WATERPROOF FINISH, T24 STRUCTURAL PLYWOOD
12. T15 CHERRY WOOD FLOORING, T24 STRUCTURAL PLYWOOD
13. T15 OAK WOOD FLOORING (HERRINGBONE), T24 STRUCTURAL PLYWOOD
14. EP, T10 CSB
15. SHIKKUI PLASTER, T10 WATERPROOF BOARD
16. T5 GLASS MIRROR
17. T15 CHERRY WOOD FLOORING, T18 STRUCTURAL PLYWOOD
18. T1.5 STAILESS STEEL VIBRATION FINISH, T12.5 GB
19. T15 TEAK WOOD FLOORING / T18 STRUCTURAL PLYWOOD
20. T70 HEAT STORAGE CONCRETE, LOOR HEATING PIPE
21. TUNNEL : T3 STAILESS STEEL VIBRATION FINISH
22. MORTAR PLASTER
23. W70 DRAIN

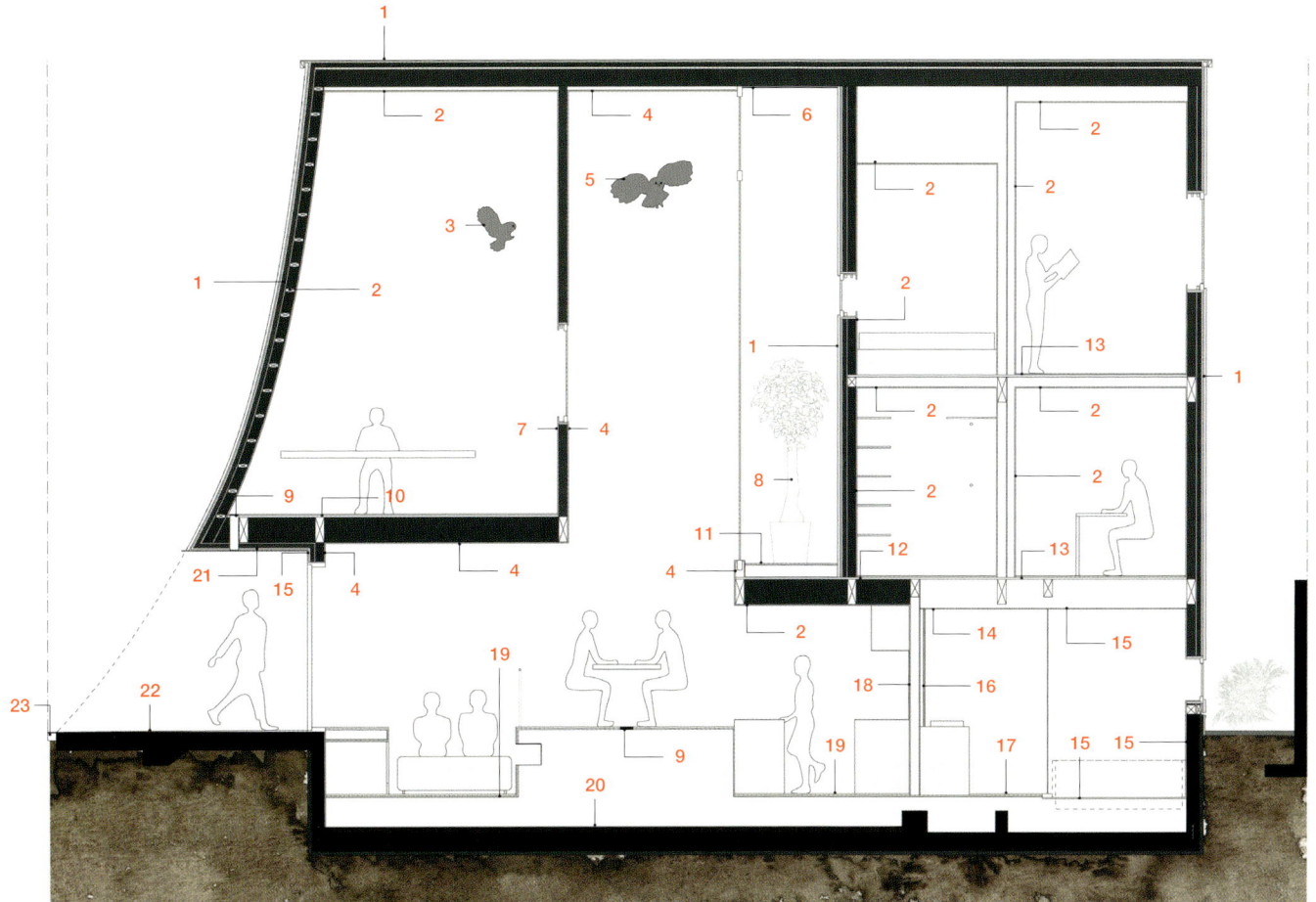

SECTION DETAIL

↑ Bookcase　↱ Light court　↳ Entrance hall

1	ENTRANCE	5	DINING	9	BATHROOM	13	STUDY ROOM
2	BIKE PARKING LOT	6	KITCHEN	10	CARTOON ATELIER	14	BED POD
3	POST STOCK	7	RESTROOM	11	LIGHT COURT	15	LIBRARY & GUESTROOM
4	SUNKEN LIVING	8	BATH ROOM	12	CLOSET		

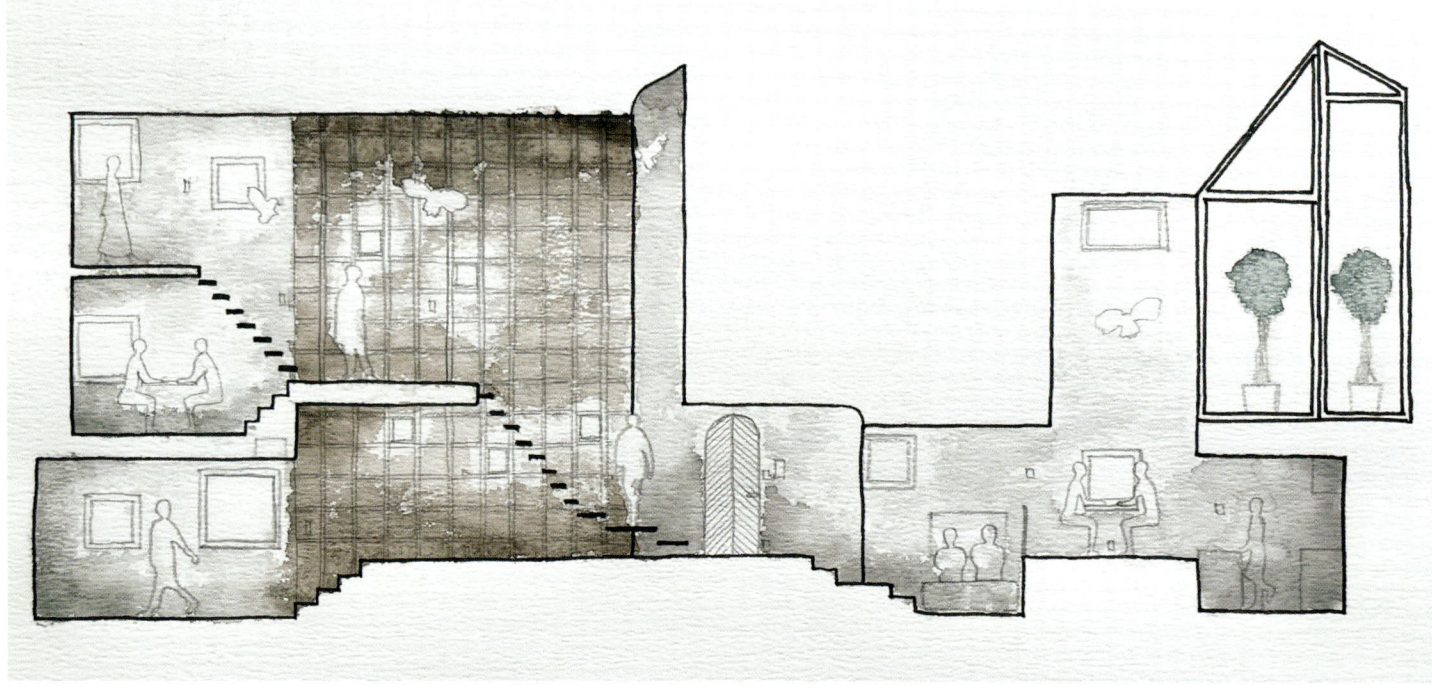

INTERIOR ELEVATION AROUND THE VOID

↱ Cartoon atelier
↳ Interior view

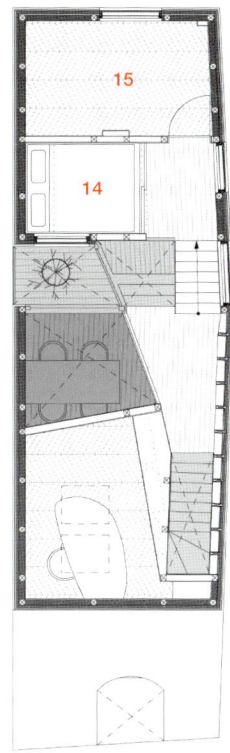

2ND FLOOR PLAN

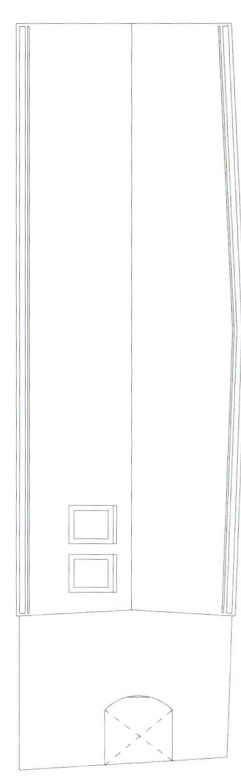

ROOF PLAN

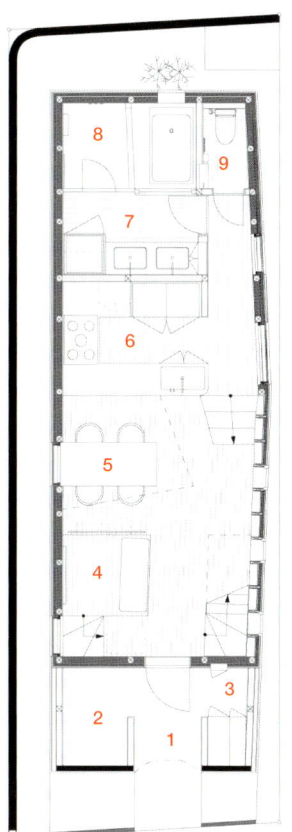

BASEMENT & 1ST FLOOR PLAN

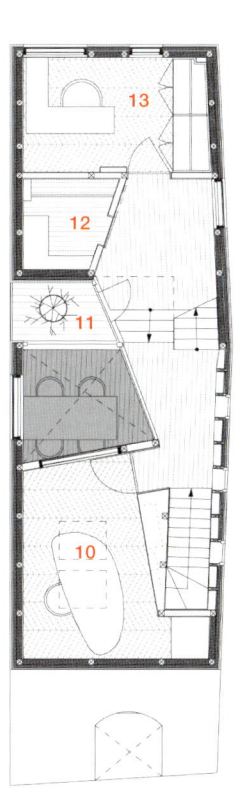

1ST & 2ND FLOOR PLAN

토양폐기물 관리오피스
OFFICE OF SOIL WASTE MANAGEMENT

ARCHITECT : SUPRA-SIMPLICITIES / HAN KUO

THIS SMALL PROJECT HUMBLY SITS IN THE COUNTRYSIDE, approximately one-hour drive from the centre of Kaohsiung city, a southern metropolis in Taiwan.

Commissioned by a client who owns a 1.27-hectare land for soil waste management and treatment, the major request was simple yet crucial that they expected this building to become the place where they could survey the whole site and be capable of immediately understanding the actual situation at any moment. This small box-like building is not only their watchhouse but also functions as an administrative office in an attempt to monitor working progress and organization of the whole area; Consequently, the design should be open without any visual obstacle to disrupt their supervision. Hence, the relationship between the office and the landscape of soil waste is the primary essence of design.

Frequently occupied by multiple types of heavy-duty "objects", including trucks and excavators, the office is depicted as one of the family members, especially as parenthood looking after other mobile equipment and vehicle. By deliberately comparing the scale of all moving objects, from biggest vehicles to smallest one, like hydraulic table cart, the design of the office is intentionally optimized in a similar scale and size.

The building consists of two levels with two distinctive characteristics, transparency and opaqueness, focusing outward and inward respectively. The first level as an administrative office is mainly a circular space enclosed by transparent glass in order to keep it visually open, giving an almost 360-degree panoramic view of the surrounding field and extending visual angle ; the second level is an opaque solid volume for the archive and storage, floating above the ground floor by stacking upon the transparent office space. It's cladded by stainless-steel corrugated sheets, neutrally reflecting the environment and sky on facade surface. Semi-outdoor shading space under the "metal box" also offers a great place to rest in the semi tropical climate of Taiwan.

Location Kaohsiung City, Taiwan **Use** Office **Site area** 2,000 m²
Built area 120m² **Gross Floor area** 240m² **Completion** 2021
Project Manager Han Kuo **Design team** Hsu NaiYuan, Lu ShinYu
Contractor Jing-Fu Co. **Photographer** Dirk Heindoerfer

이 작은 건물은 대만 남부 대도시인 가오슝시 중심부에서 차로 약 1시간 정도 떨어진 교외 지역에 있다. 토양 폐기물 관리와 처리를 위한 1.27ha의 부지 소유주는 해당 건물이 부지 전체의 상태를 언제든지 즉각적으로 파악할 수 있는 간단하지만, 중요한 역할을 원했다. 작은 상자 모양의 건물은 감시소일 뿐 아니라 작업 진행 상황과 전체 부지를 조직적으로 모니터링하는 관리 사무실이기 때문에 설계할 때 관리 감독을 방해하는 시각적 장애물 제거에 주안점을 두었다. 즉, 사무실과 토양 폐기물 부지의 관계성이 설계의 핵심이다.

트럭, 굴삭기 등 다양한 중장비 "사물" 이 주로 공간을 차지하는 사무실은 가족 구성원으로 치자면 다른 이동식 기구와 차량을 돌보는 부모의 역할을 한다. 따라서 사무실을 설계할 때 유압 테이블 카트처럼 큰 차량에서 작은 차량까지 모든 이동식 사물의 규모를 파악해 비슷한 규모와 크기로 최적화하기 위해 노력했다.

건물은 외부와 내부를 구분하기 위해 투명한 층과 불투명한 두 개의 층으로 구성했다. 1층은 관리 사무실로, 원형 공간을 따라 투명 유리창을 배치해 거의 360도의 시야각을 제공해 주변을 면밀하게 살필 수 있도록 설계했다. 2층은 보관 및 저장의 목적에 따라 불투명하고 견고한 공간으로 구성해 마치 투명한 사무 공간 위에 떠 있는 듯한 느낌을 제공한다. 골이 진 스테인리스 강판으로 마감해 외부 표면으로 하늘과 자연이 자연스럽게 반사되어 비춰 준다. "금속 상자" 아래에 빛을 차단하는 반야외 공간을 두어 아열대 기후인 대만의 더위를 피해 휴식을 취할 수 있다.

← Exterior view & entrance → Exterior view

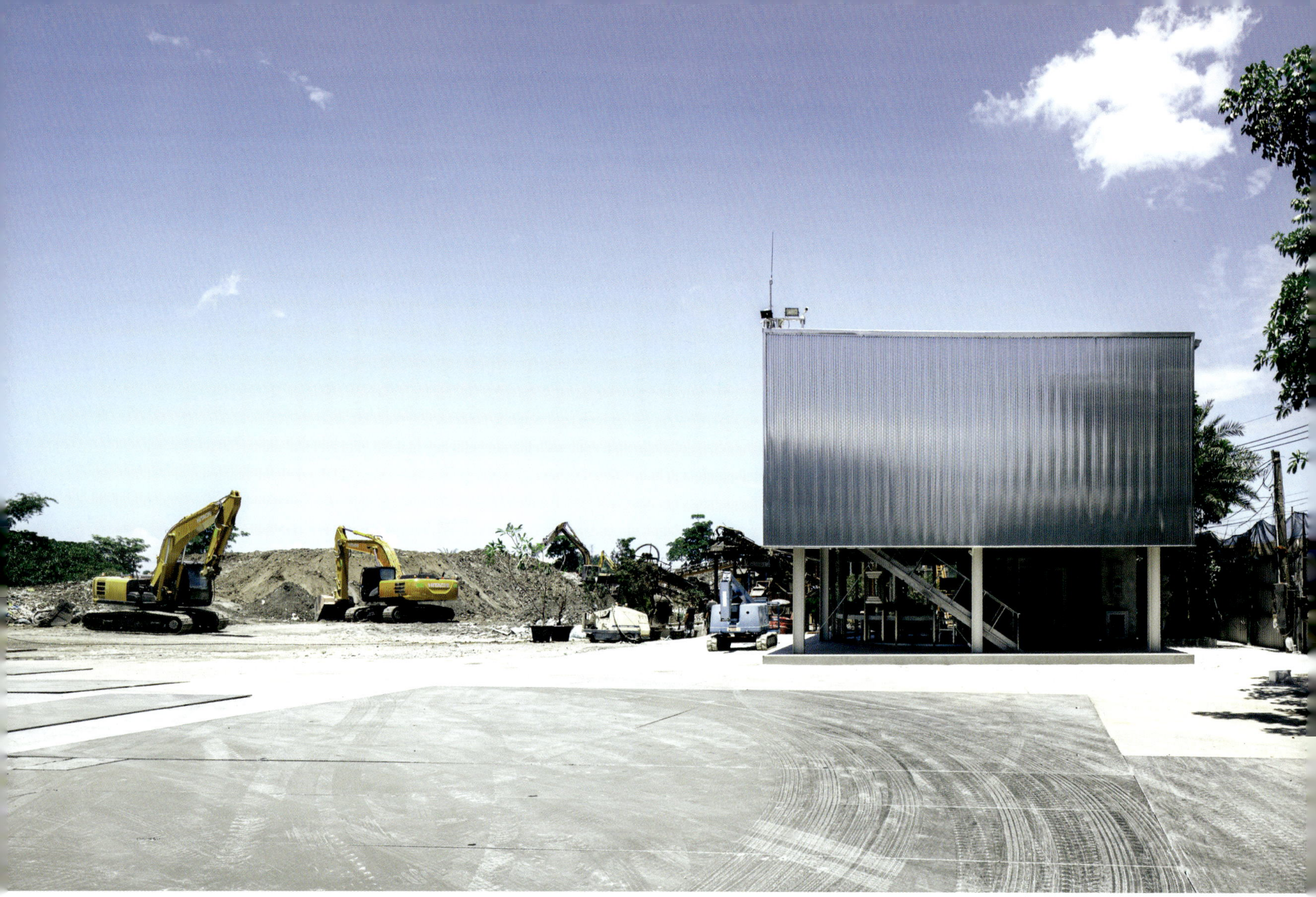

↑ The office and the landscape of soil waste ↵ The office and the landscape of soil waste → The office and the landscape of soil waste

9
8
7
6
5
4
3
2
1
0 (m)
scissor lift
boom lift
bulldoze
excavator
truck

SCALE DIAGRAM

HITACHI

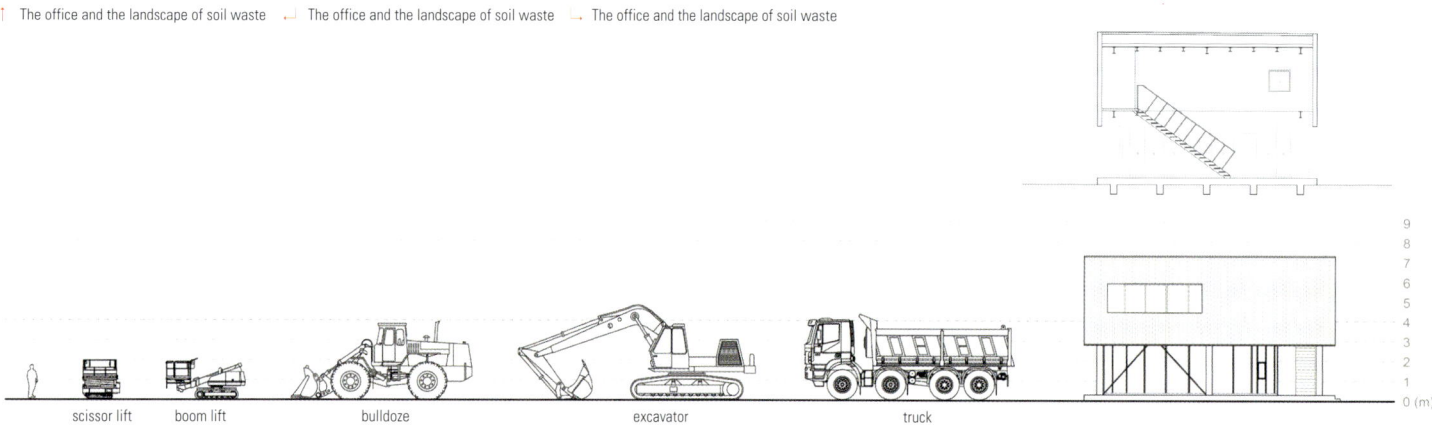

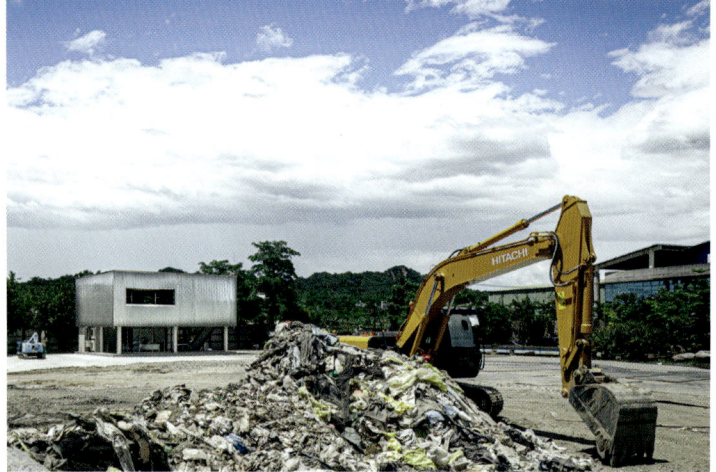

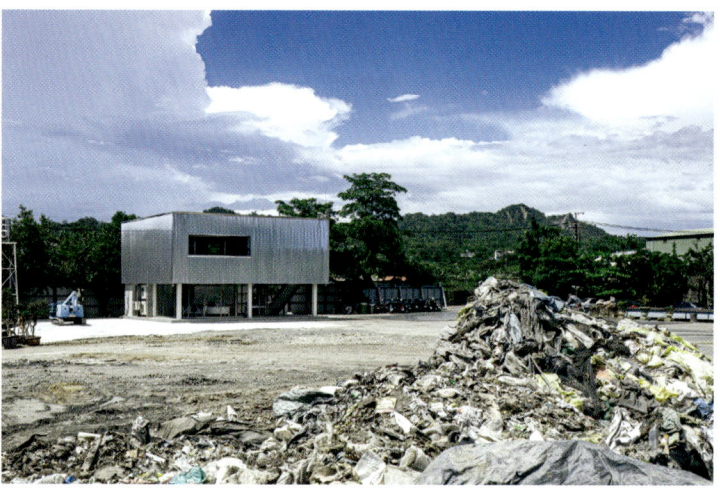

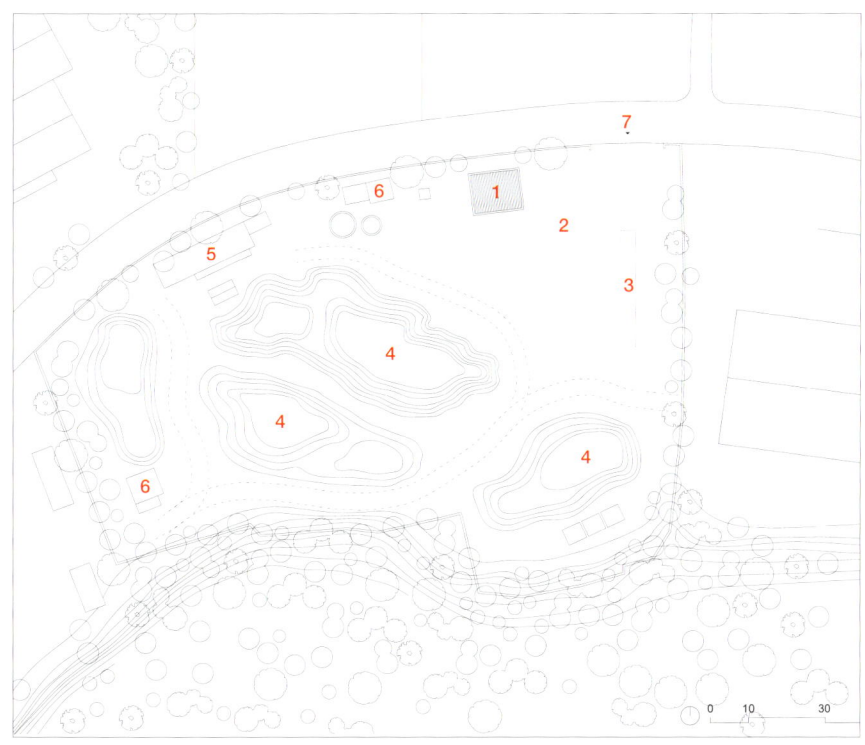

↑ The office builging night view   ↱ The office and the landscape of soil waste   ↳ The office and the landscape of soil waste

1 OFFICE	3 WASHING PLATFORM	5 SORTING AREA	7 ENTRANCE
2 PARKING	4 SOIL WASTE	6 EQUIPMENT STORAGE	

SITE PLAN

↑ 1st floor office view   ↙ Stair   → 1st floor office view

↵ Stainless-steel detail view
↓ Entrance
↳ Night office view

1 INSULATION LAYER(ROOF)
2 STAINLESS-STEEL CORRUGATED SHEET
3 POLISHED CONCRETE
4 FLUORESCENT LIGHT(4000K)
5 GIRDER AND BEAM CONNECTION
6 STEEL COLUMN BASE
7 LOW-IRON LAMINATED GLASS(8+8)
8 STRUCTURAL CONCRETE

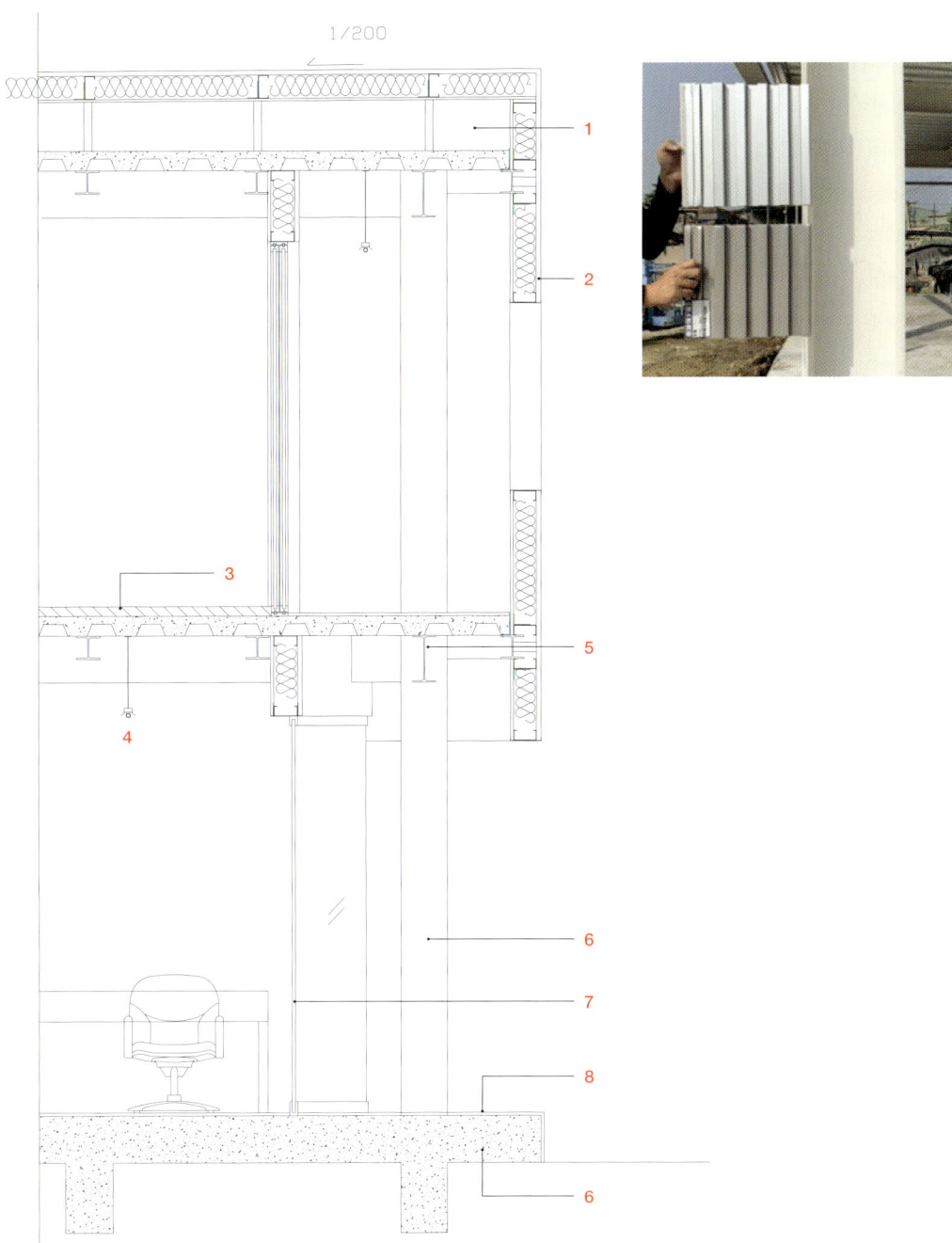

DETAIL

← Work in progress of structure system   → Work in progress of facade

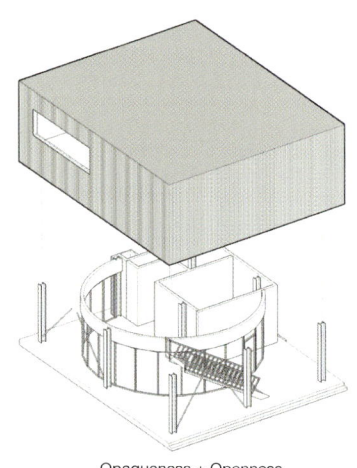 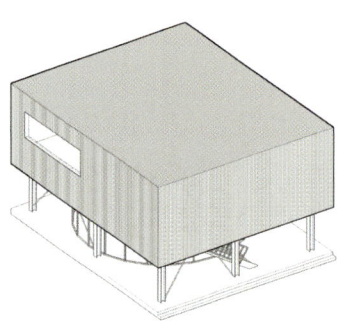 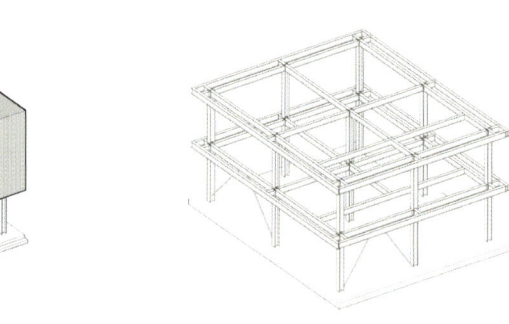

Opaqueness + Openness   Facade   Structure system

DIAGRAM

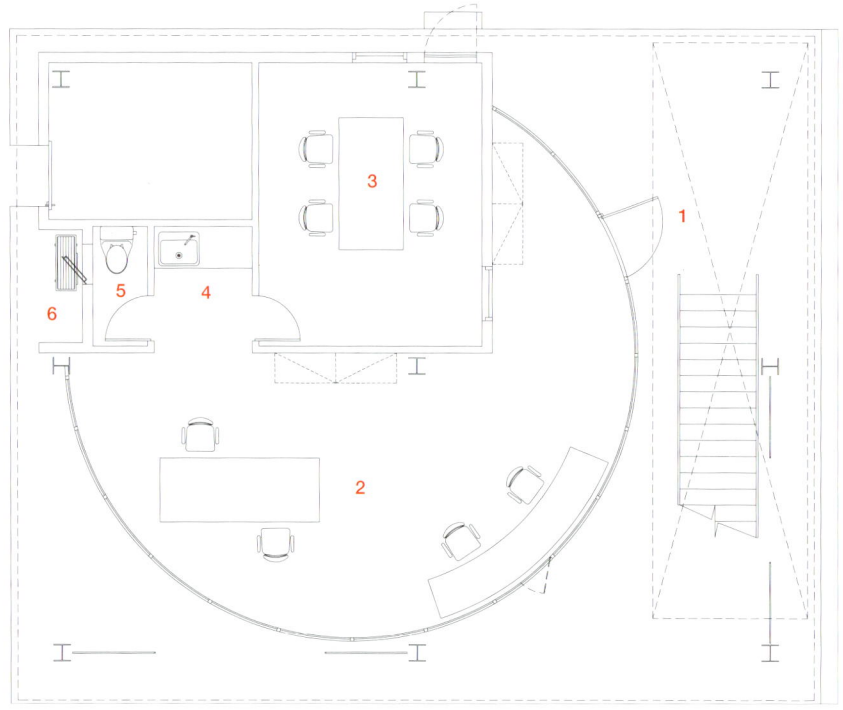

1 ENTRANCE
2 OFFICE
3 MEETING ROOM
4 PANTRY
5 TOILET
6 HVAC UNIT
7 BALCONY
8 STORAGE
9 INSULATION LAYER
10 CORRUGATED PANELS

1ST FLOOR PLAN

← Exterior night view   → 2nd floor Window

AXONOMETRIC SECTION

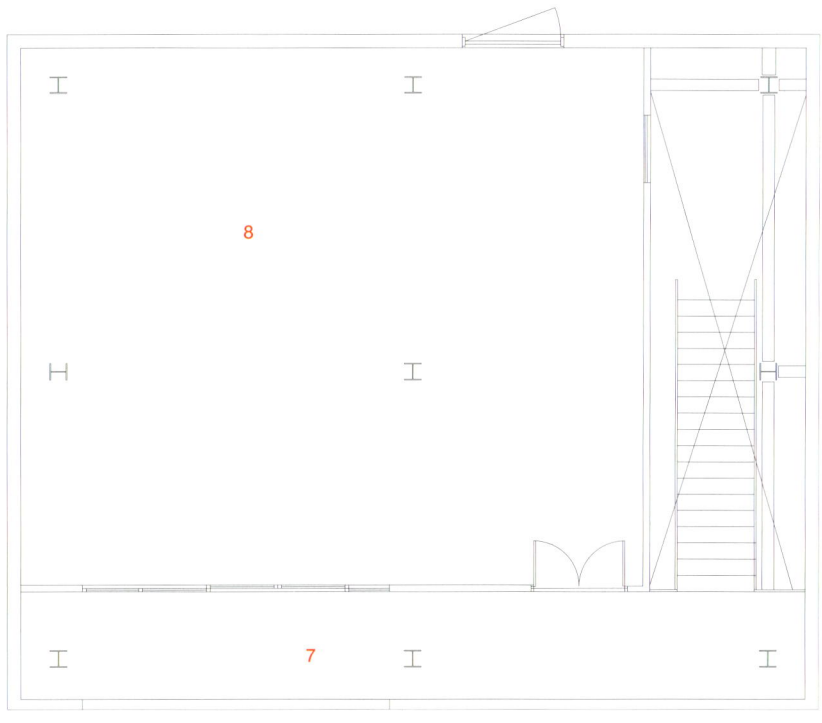

2ND FLOOR PLAN

원형광장_서클
PLAZA_CIRCLE

ARCHITECT : T2P ARCHITECTS OFFICE & P.O.T LAB / SHIKWAN YANG, TOMONORI MIURA, TATSUHITO ONO

THE ATSUGI FOREST ENVIRONMENTAL RESEARCH CENTER, a training facility owned by Kansei Kogyo, which specializes in the management and cleaning of sewer pipe networks, is located in the Morinosato district of Atsugi City, Kanagawa Prefecture, Japan.

We proposed to the client to create a plaza for visitors at the entrance to this facility, which was originally a parking lot.

Elements such as outdoor toilets, outdoor washbasins, and disaster prevention wells, which are usually located in a corner of the park, were integrated with a water garden utilizing well water and placed in the center of the site to create a highly centripetal plaza that will be the face of the company.

On the circular water garden, which is continuous with the surrounding grass, we designed a three-layered circular space that enhances the centrality of the plaza, and a different relationship with water inside and outside the circular space. The central courtyard is an outdoor washbasin space where one can feel the sound of the water flowing in the water garden. The four restrooms surrounding the courtyard, each with a view of the water garden, are quiet spaces with independent walls that block the view of the outside. On the other hand, the outside of the floating wall was planned as an active water area for children to play with water in continuity with the grass. The abstract white space that amplifies the presence of water reflects the company's continued commitment to purity.

Inspired by the sewer pipes handled by the company, three cylinder spaces of different heights were proposed. The indoor and outdoor hand wash basins were converted from hume pipes. The restroom sign was designed as a three-dimensional pictographic sign combining VP pipes, which are drainage pipes. We tried to convey the image of "pipes" as much as possible in the details of the building. We hope that visitors to the "Plaza Circle," which embodies the spirit of the company, will be inspired to think about the sewage infrastructure that supports the city, which they are usually unaware of.

Location Atsugi, Kanagawa, Japan **Use** Plaza & Outdoor toilet **Site area** 652.41m² **Building area** 24.84m² **Gross floor area** 24.84m² **Completion** 2022 **Project manager** Shikwan Yang, Tomonori Miura, Tatsuhito Ono **Design team** T2P Architects Office, Affordance, P.O.T lab **Contractor** KOJIMAGUMI **Photographer** Vincent Hecht

카나가와현 아쓰기시의 모리노사토 지구에, 하수도 관로망을 전문적으로 관리·청소하는 칸세 공업의 연수 시설 〈아쓰기의 숲 환경 리서치 센터〉의 입구에 있던 기존 주차장을, 방문객을 위한 광장으로서 계획하였다.

보통 일본의 공원의 한쪽에는 옥외 화장실이나 야외 수돗가, 방재용으로 쓰이는 우물이 분산되어 설치되는 경우가 많기에, 본 프로젝트에서는 이러한 부대시설들을 일체화해 부지 중앙에 배치함으로써, 구심성이 높은 기업의 얼굴이 되는 광장을 계획하는 것을 목표로 하였다.

주변의 잔디밭과 연속되는 원형 수공간 위에 광장으로서의 구심성을 높이는 원형 공간을 3중으로 겹쳐, 오버랩 되는 원형 공간 내·외에서 물과의 관계성을 디자인했다. 이중 중심공간에 해당되는 중정에는 물이 흐르는 소리를 느낄 수 있는 공간이며 야외 수돗가 공간이기도 하다. 이 안뜰을 둘러싸고 배치된 수공간을 바라보는 4개의 화장실은 독립 벽으로 외부의 시선이 가려진 정적인 공간이다. 다른 한편 수공간 위를 부유하는 바깥쪽 벽은 잔디밭과 연속하여 어린이들이 물놀이를 할 수 있도록 하는 동적인 공간으로 계획하였다. 물의 존재감을 증폭시키기 위해, 물결이 빛에 반사되어 공간에 표현될 수 있도록 가능한 색을 배제하였으며, 이는 청정성을 위해 노력해 온 기업의 정신이기도 하다.

기업이 취급하는 하수도관을 모티브로 높이가 다른 3개 실린더 공간과 하수도 관로의 흄관을 그대로 이용한 옥내외의 세면대, 배수관인 V관을 조합한 입체 픽토그램 사인 등 세부 디자인에 이르기까지 파이프의 이미지를 형상화하였다. 기업 정신을 구현하는 "원형광장_서클"의 방문객이 평소에는 의식하기 힘든 도시를 지탱 하는 하수도 인프라에 대해 다시 한번 생각해 보는 계기가 되기를 기대한다.

← Entrance → Night view

↑ Exterior view

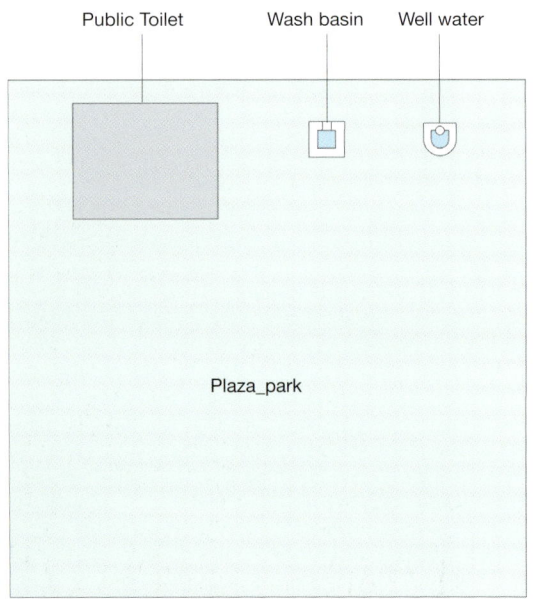
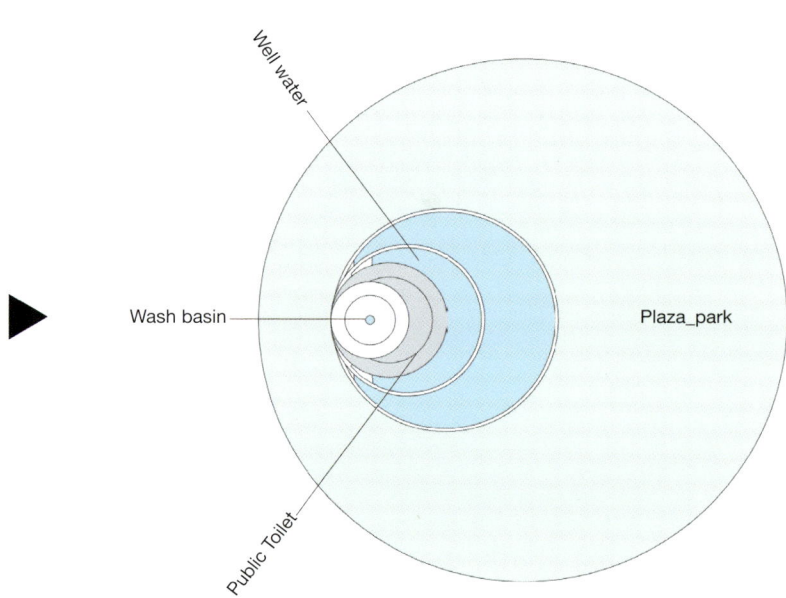

DIAGRAM

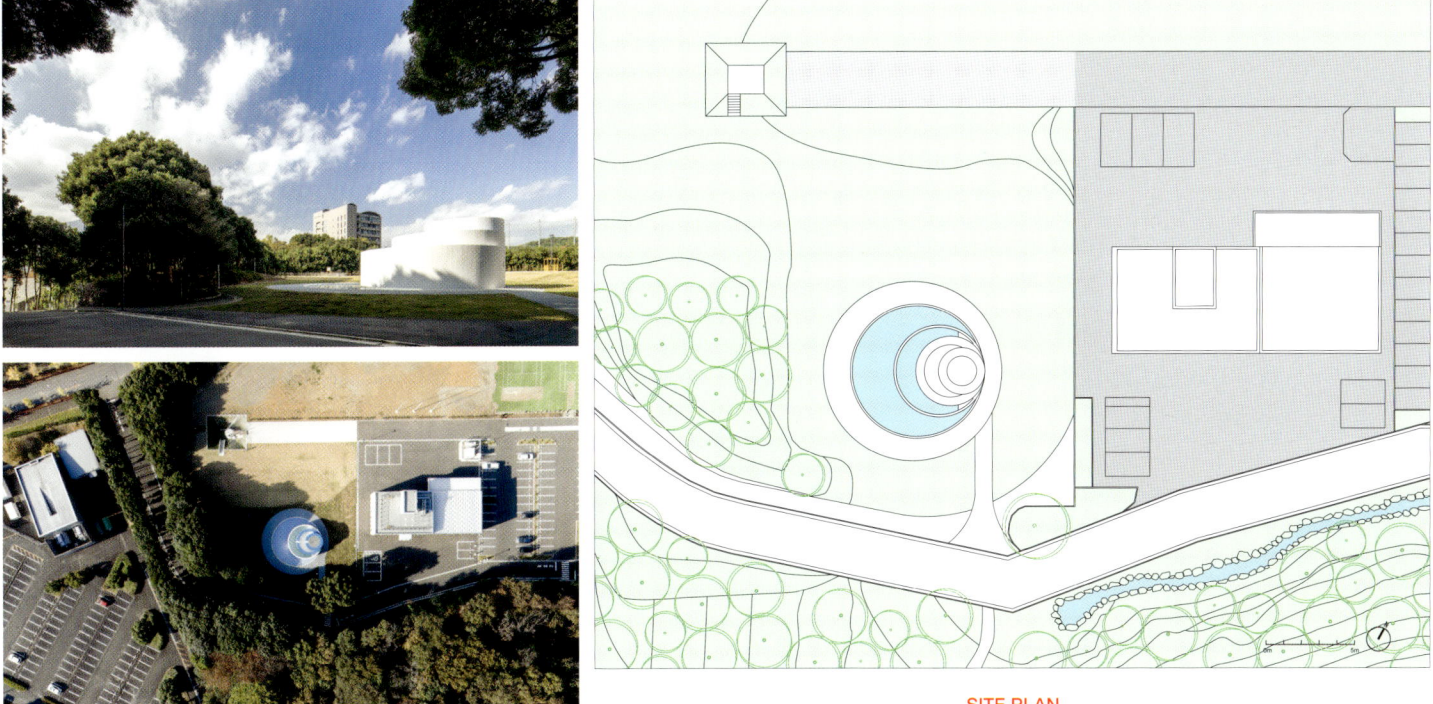

↑ View from east ← Panoramic view ↙ Bird's eyes view

SITE PLAN

← View from south   → View from south   ↱ Bird's eyes view   ↳ Night view

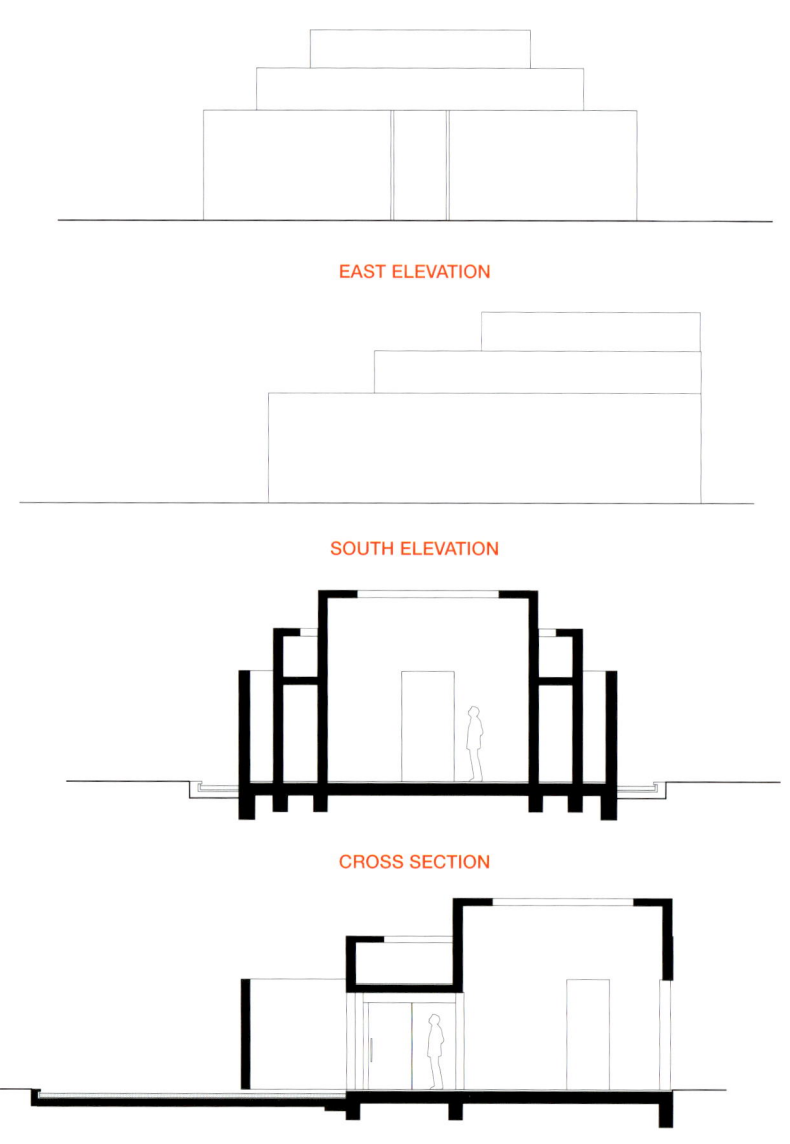

EAST ELEVATION

SOUTH ELEVATION

CROSS SECTION

LONGITUDINAL SECTION

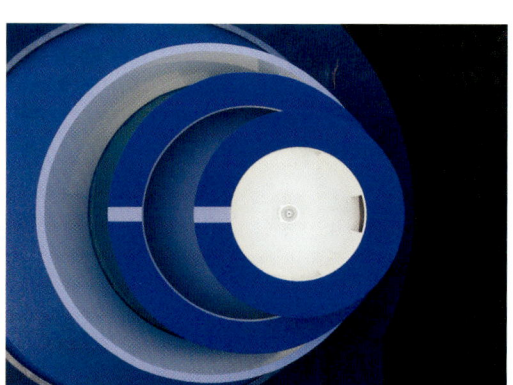

1 CONCRETE FINISH	4 WATER LEVELING PAINTED IN N90	7 SHEET WATERPROOF	10 LIGHTING CUT
2 Ø19 MOSAIC TILE	5 PAINTED IN KANSEI COMPANY CI COLOR	8 MORTAR	11 Ø19 MOSAIC TILE (WHITE)
3 CONCRETE FINISH PAINTED IN N90	6 AL WATERING	9 WATERING JOINT	12 GUTTER : PEBBLE

SECTION DETAIL

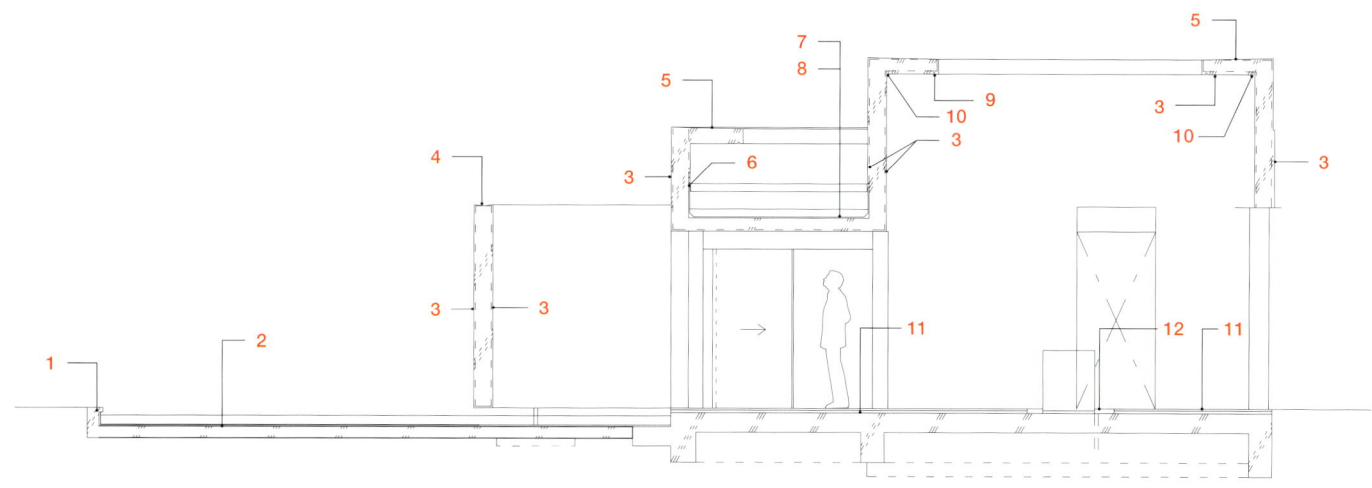

← Water garden   → Water garden

239

↑ Courtyard   ↙ Courtyard   ↘ Courtyard

← Hallway   → Toilet

1 WATER GARDEN
2 COURTYARD
3 MEN TOILET
4 INFANT-FRIENDLY TOILET
5 ACCESSIBLE TOILET
6 WOMEN TOILET

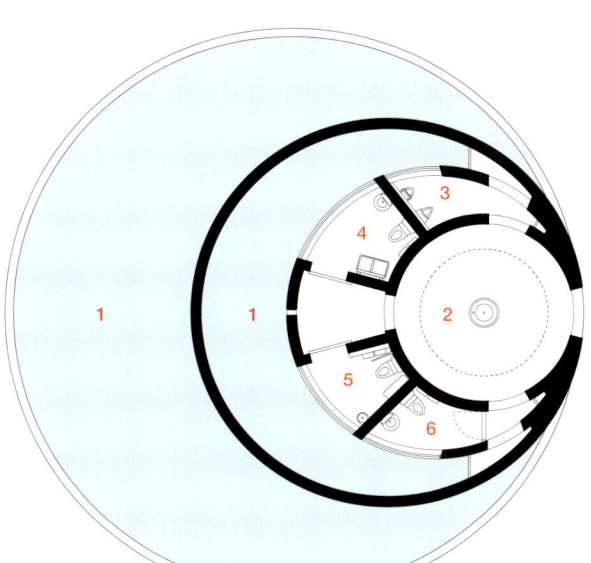

1ST FLOOR PLAN

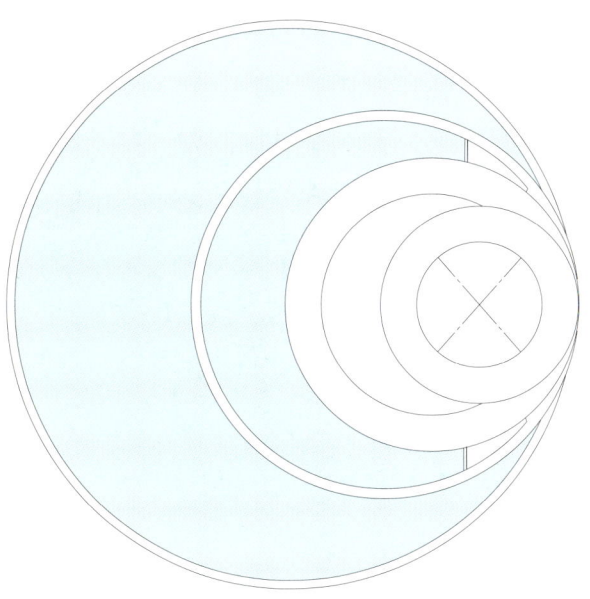

ROOF PLAN

푸하이 부동산 빌딩

PU-HAI PROPERTIES BUILDING

ARCHITECT : AOMO / SIVICHAI UDOMVORANUN

PU-HAI OFFICE BUILDING, A 4-STORY OFFICE BUILDING. We locate the building to make a way to coexist between buildings and trees. The project is located on Sanseab canal and surrounded by many offices buildings. Most of them are located on high-rise building and are enclosure area. What want to make our project difference and provide another alternative for work environment. We want to create a community of workplaces to make the users enjoy their work life more. We designed to divide the buildings in to 3 parts to let the natural ventilation and light inside more and to minimize the building scale as well. This also brings natural light into the office space as well to reduce the use of electricity in the daytime.

The project is composed of 3 buildings. 2 building are office space and the other is service core for stairs, elevator, WC and all mechanical rooms. The Service building which is more solid we put them against the West side to shield the project from the Sun. For the building structure, we decided to use light weight construction to save construction time and also to avoid the site damage to all those trees. The ground floor we choose concrete column to reduce damage that might occurs from vehicles and flood. Steel beams and columns are used for the rest of the building except for the service building which is more appropriate to use typical RC structure. We also use pre-insulated metal panel for both roof and walls to minimize the masonry works that can damage the site and the trees and for faster construction time.

Center court with existing trees connects all the floors together and make people enjoy walking on the stairs than using the elevators and also can interact to each other more. Although having a lot of opening is better for natural light but also taking strong glare and heat gain. The position of opening is very important. We use expanded metal sheet awning over building windows to filter the sunlight according to the sun direction.

Location 1974/1 New Phetchaburi Road, Bangkapi, Huay Khawng, Bangkok, Tailand **Use** Offices **Site area** 1,100m² **Built area** 1,980m² **Gross Floor area** 2,294m² **Completion** 2021 **Project Manager** Sivichai Udomvoranun **Design team** Kumpee Charoensook **Contractor** Tt Construction Co., Ltd. **Photographer** DOF Sky Ground

4층 규모의 푸하이 오피스 빌딩은 건물과 나무가 함께 어우러질 수 있도록 설계되었다. 건물은 산샙 운하에 있으며, 수많은 사무실 건물에 둘러싸여 있다. 사무실 대부분은 담장을 올린 지역의 고층 건물에 있다. 우리는 다른 사무 환경과는 구별되는 대안 공간을 만들고자 계획했다. 직장인들이 회사에서 더욱 즐겁게 일할 수 있도록 직장 내 커뮤니티를 만들고자 했다. 건물을 세 영역으로 나누어 내부의 자연적인 환기 및 채광 기능을 향상하고 건물의 규모도 최소화했다. 결과적으로 사무실에 자연광을 더 잘 들어오게 하여 주간 전기 사용량을 줄일 수 있었다.

이 프로젝트는 세 개의 건물로 구성되어 있다. 두 개의 건물은 사무실 공간과 계단, 엘리베이터, 화장실 및 기계실 등 기타 공간으로 구성되어 있다. 한층 견고한 서비스 건물은 직사광선을 차단하기 위해 서향으로 배치했다. 건물 구조는 공사 시간을 절약하고 부지의 나무 손상을 방지하기 위해 경량 구조로 결정했다. 1층에는 차량 및 홍수로 인한 피해를 최소화하고자 콘크리트 기둥을 세웠다. 일반적인 RC 구조가 적합한 서비스 건물을 제외한 나머지 건물에는 철재와 기둥을 사용했다. 또한 지붕과 벽체에는 단열 처리한 메탈 패널을 사용해 부지와 나무에 손해를 입힐 수 있는 석조 공사를 최소화하며 공사 시간을 단축했다.

기존에 나무가 많은 중앙 지역은 모든 층으로 연결되어 사람들이 엘리베이터를 타는 것보다 계단으로 올라가며 서로 상호 교류를 할 수 있도록 설계했다. 개폐된 곳이 많을수록 자연 채광에는 좋으나 동시에 눈부심과 열이 많이 발생할 수 있다. 그래서 개폐 위치가 매우 중요하다. 건물 창문에 태양 방향에 따라 확장된 금속판 차양을 더해 직사광선을 차단했다.

← Exterior view   ↙ Exterior night view

↑ Building entrance

ELEVATION

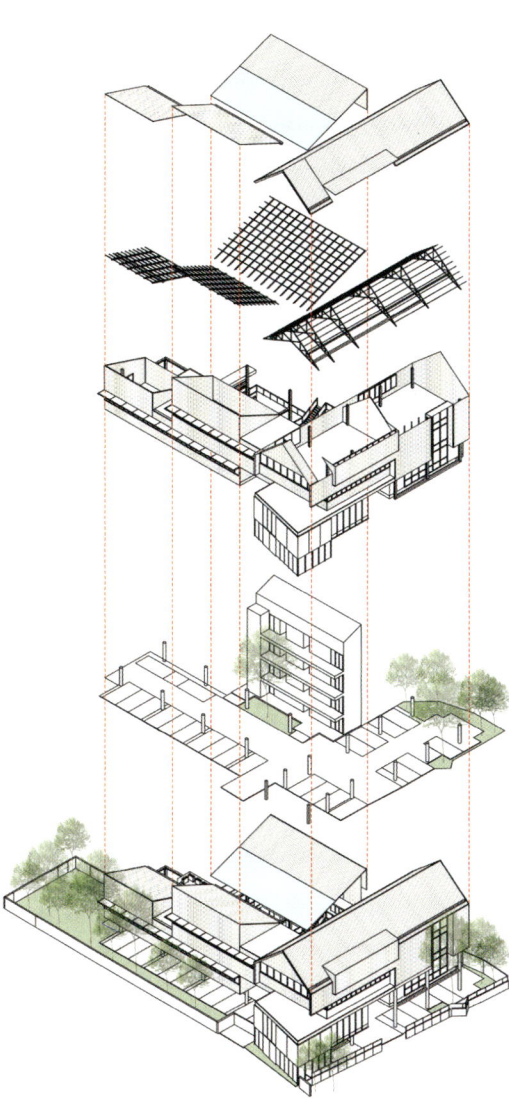

ROOFING MATERIAL
Insulated 2" PU Metal sheet panels

ROOF STRUCTURE
Steel frame beams and Purlins and
Light-gauge rafters

WALL
Insulated 1" PU Metal sheet panels
on 2"X4" steel frame
w/9mm Gymsum interior wall board

BUILING STRUCTURE
FLOOR(2ND-4TH FI)
Steel post and beam frame
w/ precasted concrete plank
w/ precasted cement floor finish

GROUND DLOOR AND SERVICE
BUILDING
RC Structure
RC Column and RC floor
w/ masonry wall, platered and painted

OVERALL BUILDING

CONSTRUCTION ISOMETRIC

↑ Night exterior view ↓ Main street view

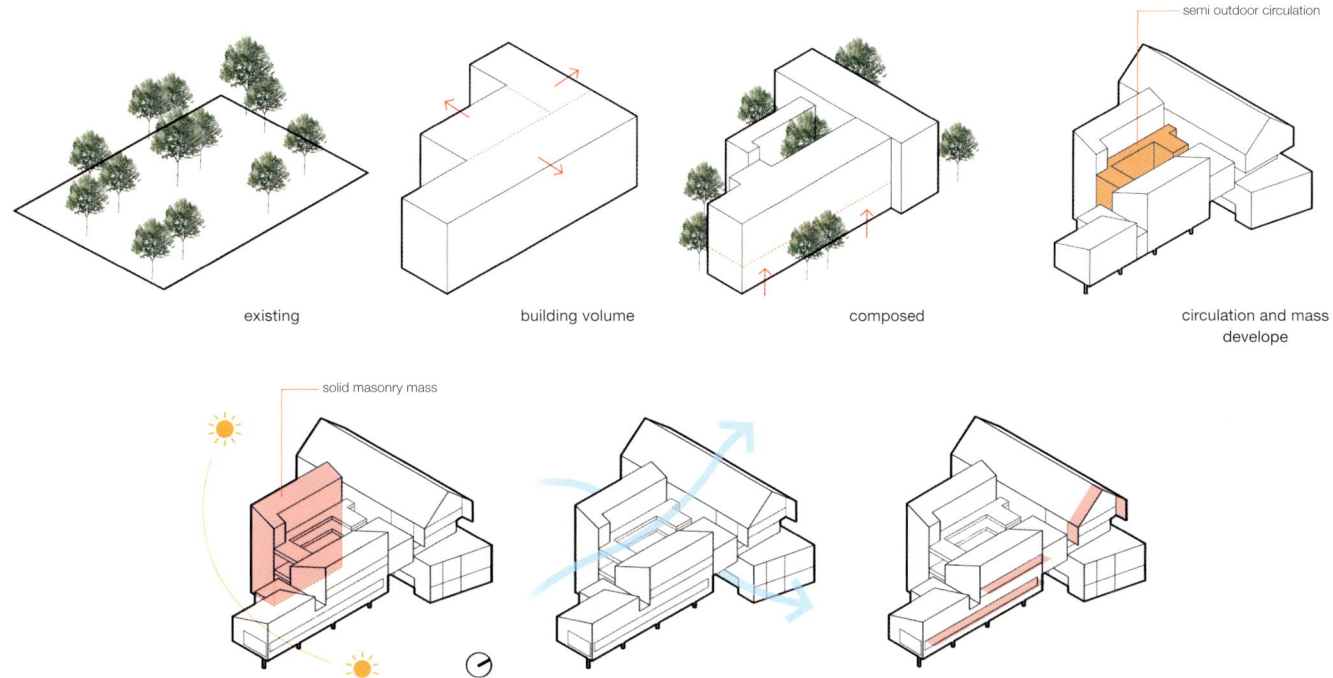

existing   building volume   composed   circulation and mass develope

Put service building against the West side to shidld the project from the Sun   ventilation   east side sun protect

DESIGN DEVELOPMENT DIAGRAM

245

↑ Steel beams and columns structure

1 OFFICE
2 PARKING LOT
3 TOILET
4 WATER TANK

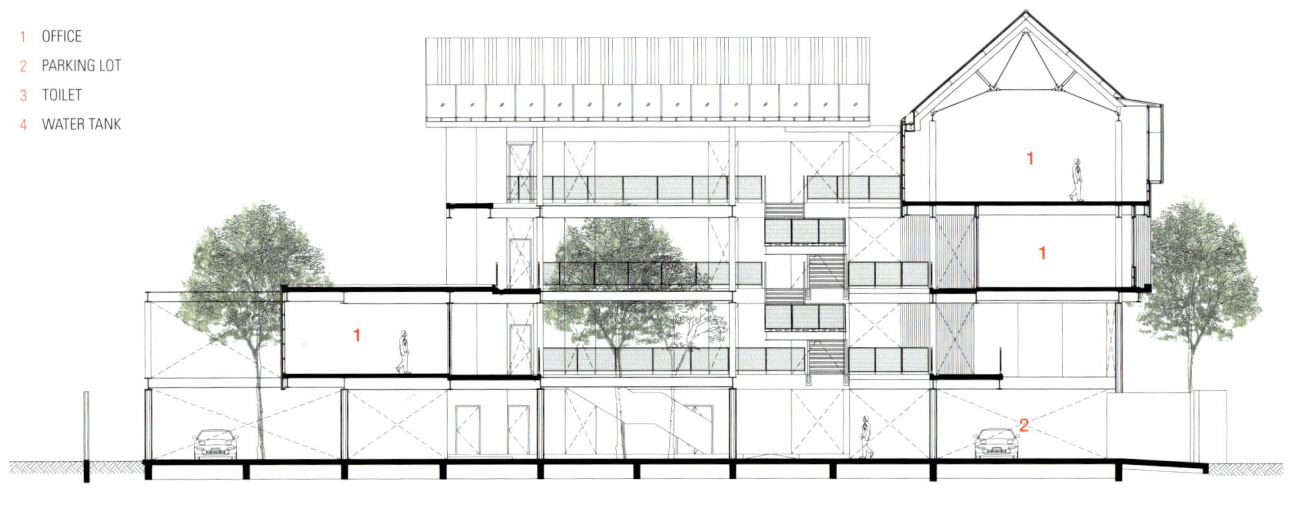

SECTION

→ Existing trees connects all the floors together and make people enjoy walking on the stairs

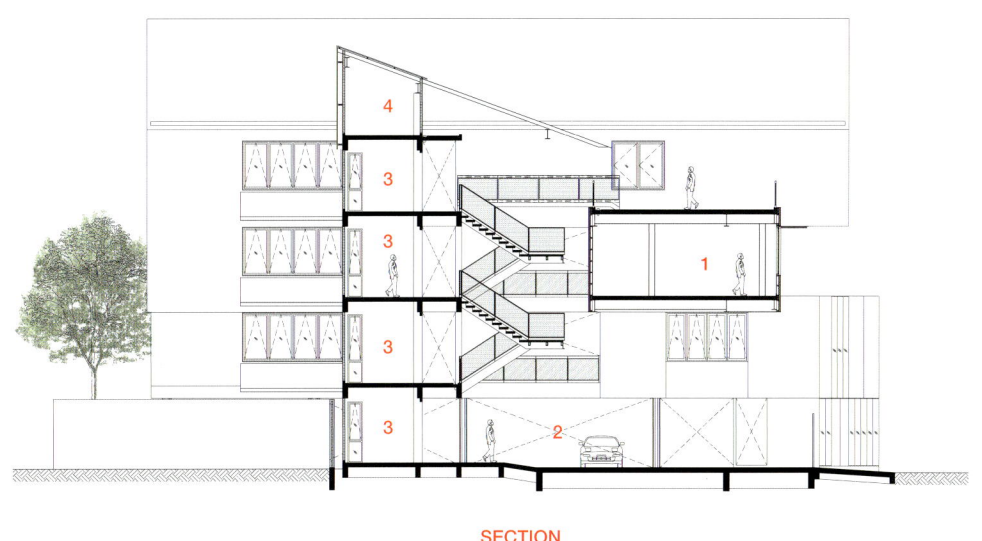

SECTION

↑ Center court with existing trees　↙ Coexist design between buildings and trees　↳ Main stair & tree

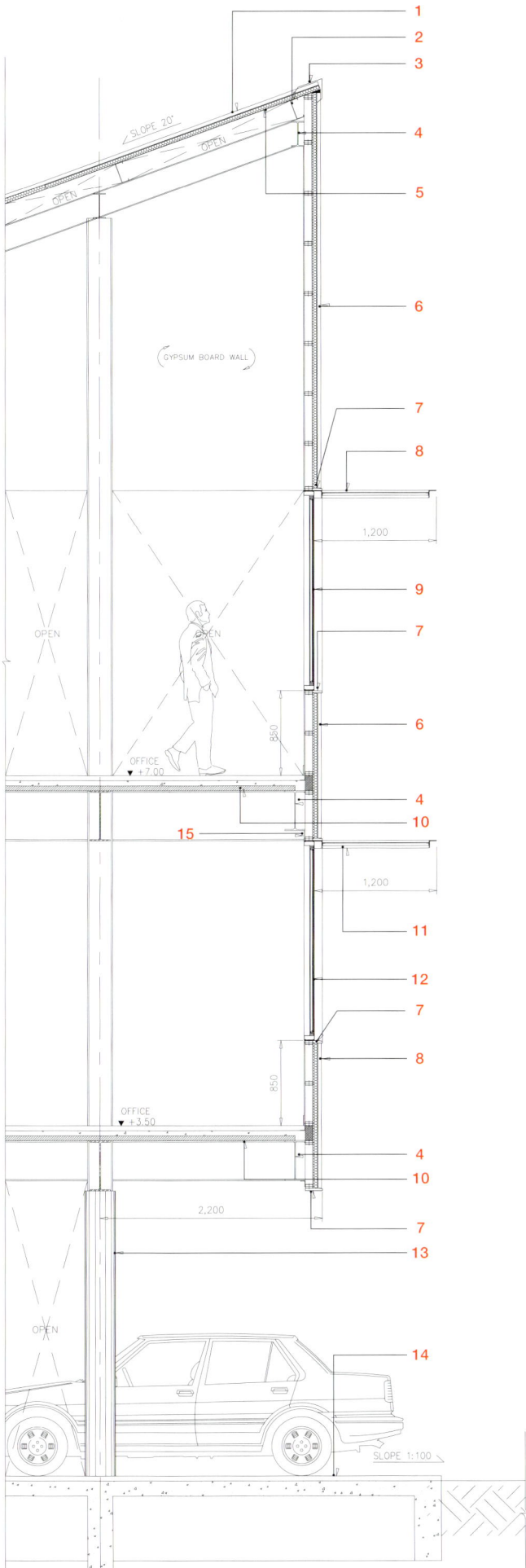

1. METAL SHEET ROOF
2. LIGHT GAUGE RAFTER
3. METAL ROOF FLASHING
4. STEEL BEAM
5. 2" PU FOAM INSULATED METAL SHEET
6. 1" PU FORM INSULATED METAL SHEET 1.5"X3" STEEL FRAME 9mm GYPSUM BOARD
7. METAL FLASHING
8. EXPANDED METAL PANEL 3"X3" STEEL ANGLE FRAME
9. ALUMINIUM WINDOW FRAME
10. 50MM CEMENT FLOOR PRECASTED RC. FLOOR PLANK
11. EXPANDED METAL PANEL 3"X3" STEEL ANGLE FRAME
12. ALUMINIUM WINDOW FRAM
13. EXPOSED RC. COLUMN
14. FLOOR HARDENER
15. GYPSUM BOARD WALL

WALL SECTION DETAIL

↑ Existing trees connects all the floors   ↓ Natural light into the building space

← Center court with existing trees   → Existing trees connects all the floors and make people enjoy walking on the stairs

2ND FLOOR PLAN

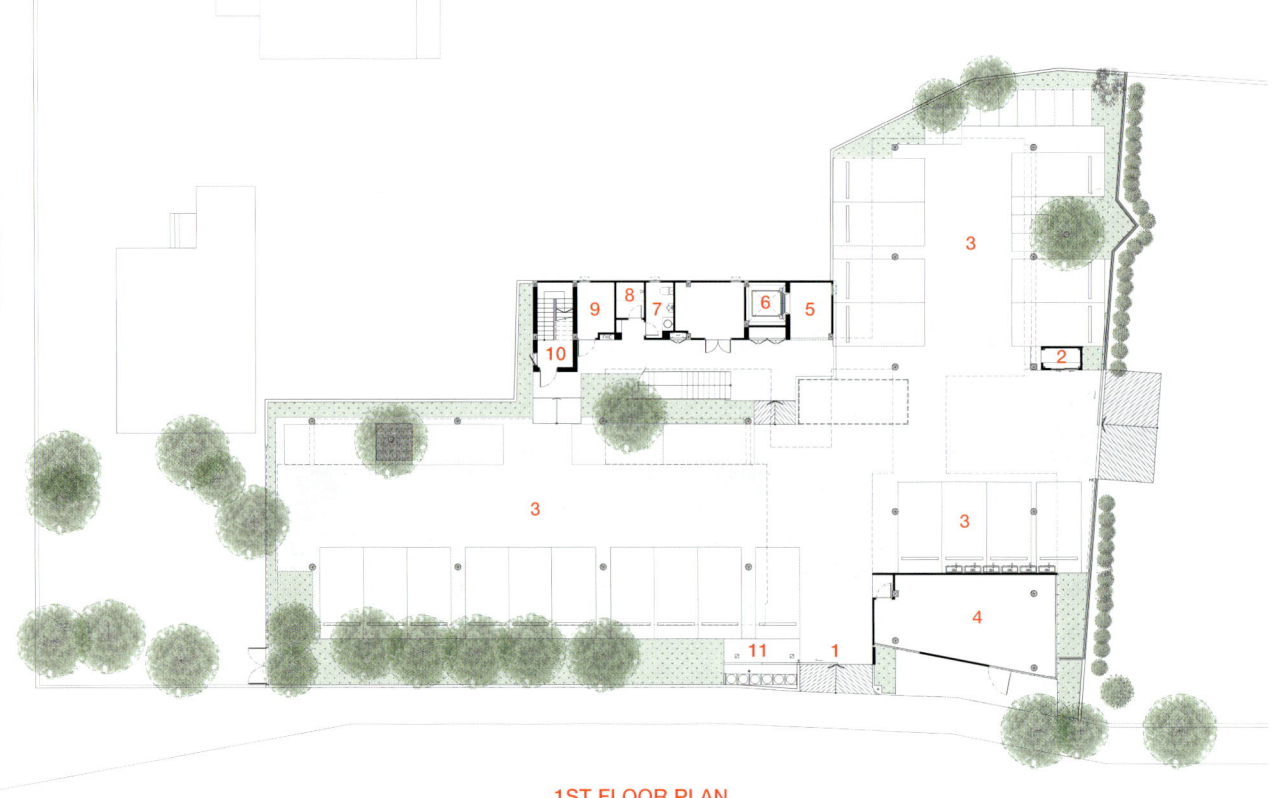

1ST FLOOR PLAN

1 ENTRANCE
2 GUARDHOUSE
3 PARKING AREA
4 ROOM FOR RENT
5 FOYER
6 LIFT
7 TOILET
8 BATHROOM
9 ELECTRIC ROOM
10 FIRE ESCAPE
11 GARBAGE
12 PANTRY
13 CORRIDOR
14 SERVICE AREA

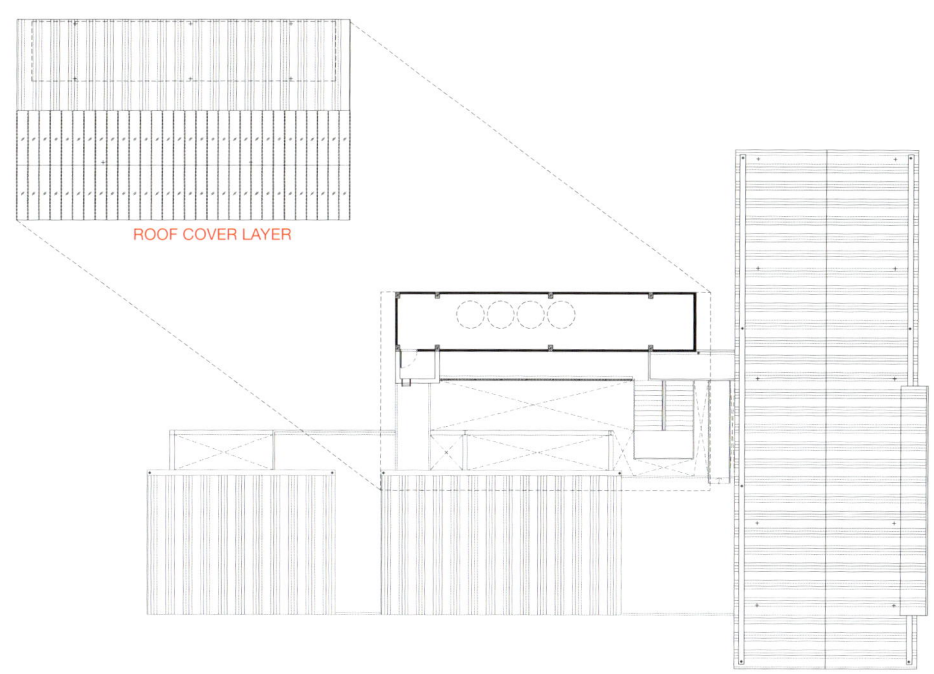

ROOF FLOOR PLAN

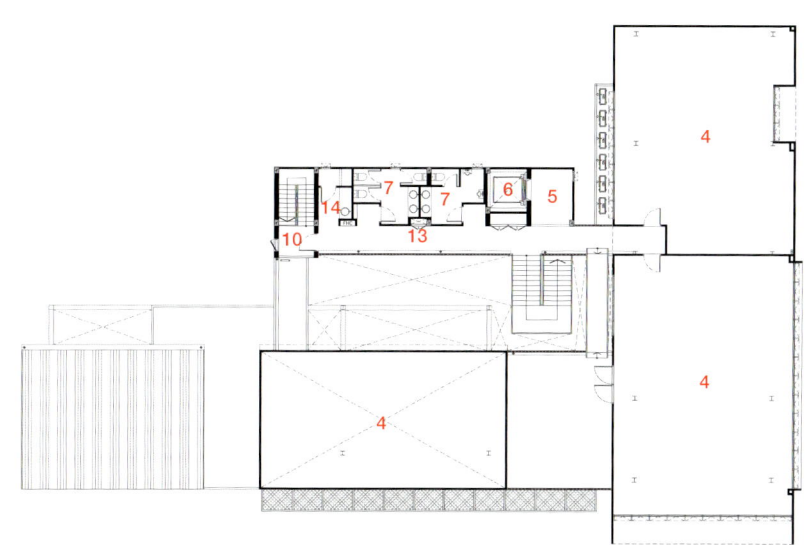

4TH FLOOR PLAN

↑ Stair   ↓ Stair

3RD FLOOR PLAN

251

다윈 1111
DARWIN 1111

ARCHITECT : HERMANOS GOLDENBERG / JAVIER GOLDENBERG, MATIAS GOLDENBERG

WHEN WE WERE ENTRUSTED WITH THE PROJECT FOR A RESIDENTIAL BUILDING, with parking and a commercial space for gastronomy at the corner of Darwin and Loyola streets, in front of an old tannery from the last century that was repurposed as lofts several decades ago, by the regulations of construction in force at the time of receiving the order, the buildability factor was less than double the surface of the land, which led us to think of proposing two volumes joined by a circulation bridge separated by a street or interior patio, a typology taken precisely from the tannery to which I was referring, a building that we know a lot about because many colleagues, designers, photographers and creatives have their studios there. Due to our experience in adding value to protected patrimonial buildings, the use of steel structures is present in our projects and we had already built a new photography studio using W steel beams with collaborating steel deck. With the factor of buildability, the typology and the program, the choice of the construction method was undoubted and as we worked with our engineering consultants and suppliers, the project took on volume and functionality almost naturally, in an interdisciplinary design process that took practically one year. The level of precision and detail necessary to manufacture a building in an industrial plant, bring it disassembled and assemble it in four stages every 45 days requires meticulous and dedicated project, industrial, logistics and material coordination. Prior to assembly, the columns were excavated, submerged, and founded at a depth of three meters below plot level for a subsoil that contains 11 garages accessible by a hydraulic lift and technical rooms for pressurized water supply, electric meters, and services panels. The first stage of assembly included the slab of precast joists on the basement, the second on the ground floor, the third on the first and the fourth on mezzanines between the double heights generated by these levels. To complete the reinforcement of the mezzanines, a compression layer of reinforced concrete was made with 8mm electro-welded mesh and mechanically flamed with a helicopter, which forms the final pavement of the units. The dividing walls between interiors and exteriors were then raised with alveolar concrete bricks finished with two- or three-layer sprayable render. Next we began the assembly of the PVC carpentry and the assembly of the glass brick walls to complete the exterior enclosure, because in functionality we consider that these bricks generate intimacy and special lighting which also helped us to rationalize the design of the windows reducing them to a few typologies that were used in all the units. Once the building was closed, the interior and exterior access stairs to the terraces were assembled, together with the interior and exterior railings and TDL mesh-type trellis. The reinforced concrete staircase, hung with angular tensioners at the landings, was made as a suspended beam that gradually adapts to the angle generated by the implantation of the volumes on Darwin and Loyola.

Location Ciudad Autonoma de Buenos Aires, Argentina **Use** Housing **Gross floor area** 1,295.38m² **Completion** 2022 **Project manager** Javier Goldenberg **Photographer** Federico Kulekdjian

← View at sunset → Partly view of exterior

↑ Exterior view

수십 년 전 다락방으로 용도 변경된 오랜 무두질 공장 앞 다윈 거리와 로욜라 거리 모퉁이 대지에 주차장과 식음 공간을 갖춘 주상복합건물을 건설하는 프로젝트가 맡겨졌을 때, 당시 시행 중인 건축 규정상 시공성 요소가 대지면적의 2배도 되지 않거나 내부 파티오로 분리되는 원형 다리를 연결해 잇는 두 개의 구조를 제안하기로 하였다. 이는 건축가의 동료, 디자이너, 사진작가 및 창작자들의 스튜디오가 많이 위치해 있어 익숙한 동시에 앞서 언급된 무두질 공장에서 영감을 얻은 디자인이다. 잘 보존되어 온 건물을 더욱 가치있게 만들기 위해서 철골 구조를 선택했으며, W 강철 빔과 강철 데크를 활용한 사진 스튜디오를 새롭게 구성하였다. 시공성 요소와 건물의 유형은 시공법을 결정하게 되었는데, 엔지니어링 컨설턴트 및 계약 업체와 함께 작업하면서 거의 1년이 걸리는 학제 간 설계 프로세스를 통해 프로젝트의 규모와 기능이 자연스럽게 이루어졌다. 산업 공장에서 건물을 사전 제작하기 위해 45일마다 4단계로 분해 및 조립하는 단계에서 필요한 정밀하고 전문적인 프로젝트, 산업, 물류 및 자재 관리 등의 정밀함과 세심함이 담겼다. 수압 리프트로 접근할 수 있는 11개의 차고와 가압수 공급, 전기 계량기 및 서비스 패널을 갖춘 기술실을 위한 하층토를 확보하기 위해 조립하기 전에 도면상의 지반 3m 아래 깊이를 굴착해 기둥을 세우게 되었다. 사전에 성형된 조이스트 슬래브를 조립의 첫 번째 단계에서는 지하실, 두 번째에서는 지상층, 세 번째에서는 1층, 네 번째에서는 중이층에 설치했다. 중이층을 보강하기 위해 전기 용접한 8mm 메쉬로 콘크리트 압축 층을 보강하고 헬리콥터를 이용해 기계적으로 열을 발생시켜 해당 부분을 마감 처리하였다. 내부와 외부를 분리하는 벽의 높이는 렌드를 두세 겹으로 뿌려 마감한 다공성 콘크리트 벽돌로 올렸다. 다음으로는 PVC 목공 작업 및 벽돌 형태의 유리 벽을 조립해 외벽을 완성했는데, 그 이유는 이 벽이 기능 면에서 익숙하면서도 특수 조명을 활용해 자칫 평범해 보일 수 있는 창 디자인을 한층 새롭게 연출할 수 있기 때문이다. 건물이 폐쇄된 후, 내부 및 외부 철책과 TDL 메쉬 격자 구조물을 이용해 테라스로 이어지는 내부와 외부 계단을 조립했다. 계단참에 각진 텐셔너를 단 철근 콘크리트 계단은 다윈과 로욜라에 세워진 구조물의 각도와 점차 맞춰지도록 빔을 매달아 완성했다.

↑ Front facade   ↳ Exterior view

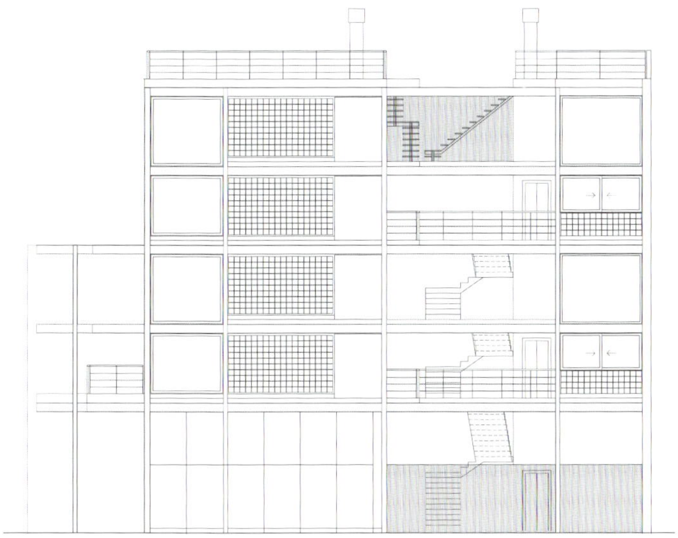

FRONT ELEVATION

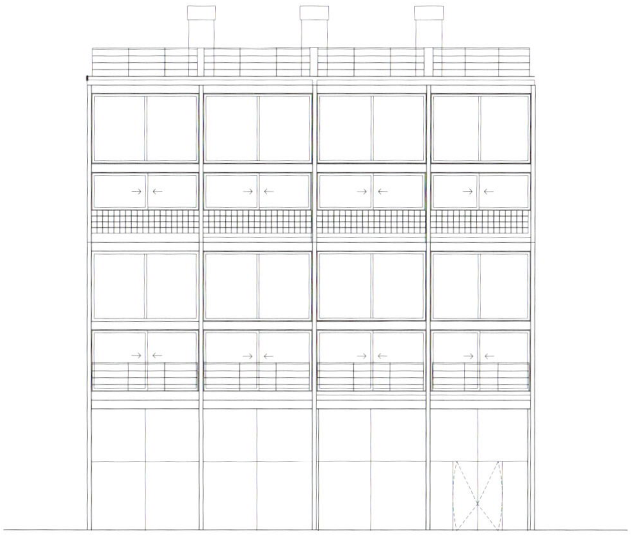

SIDE ELEVATION

Courtyard

↖ Courtyard ↑ Exterior view ↙ View from street

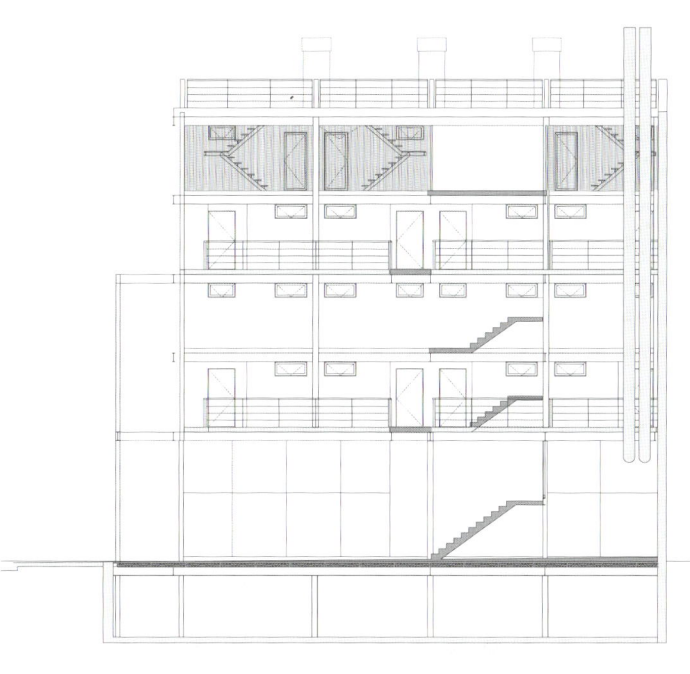

SECTION

↑ Interior view　← A two-story studio　↱ Kitchen　↳ Rooftop terrace

1 COMMERCIAL UNIT
2 EXTERIOR STAIRCASE
3 CORRIDOR
4 RESIDENCE UNIT
5 MEZZANINE FLOOR
6 BATHROOM
7 ROOFTOP

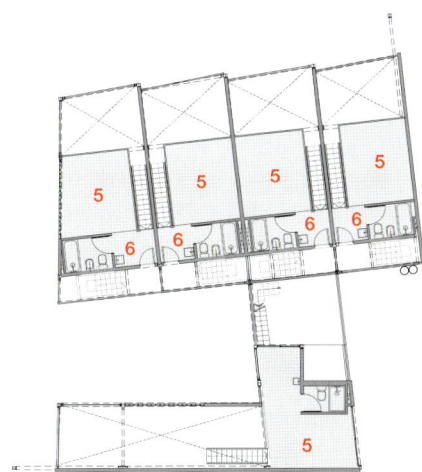

2ND MEZZANINE FLOOR PLAN

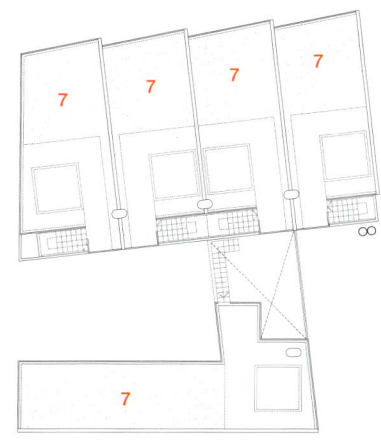

ROOF FLOOR PLAN

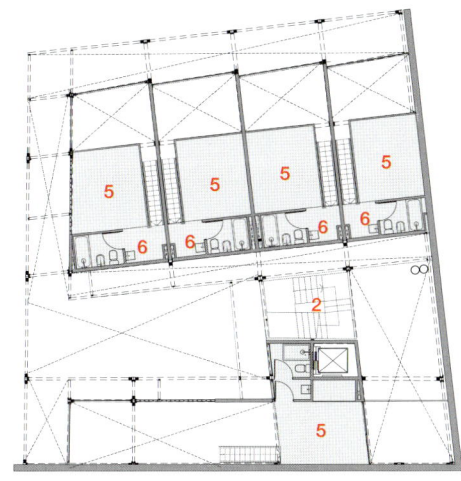

1ST MEZZANINE FLOOR PLAN

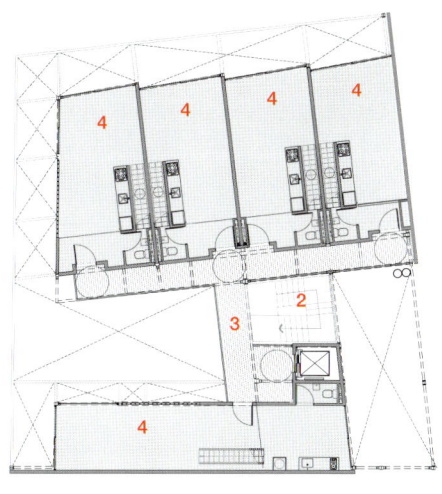

2ND FLOOR PLAN

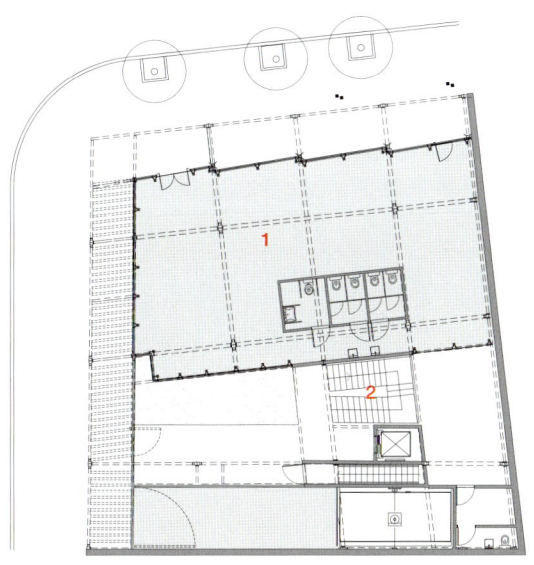

GROUND FLOOR PLAN

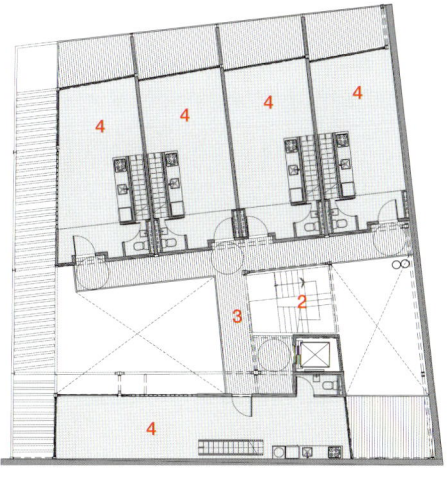

1ST FLOOR PLAN

알에이치플러스 빌딩

RH+ BUILDING

ARCHITECT : RBK ARQUITECTURA / MARCELO REBECCA, LUCIANO REBECCA, MATIAS REBECCA, VICTORIA POLLETI

THE RH+ PROJECT WAS BORN, SEEKING TO RECOMPOSE THE URBAN SHORTCOMINGS and alter them to improve the quality of life of its occupants. In a lot that was ninety-five percent occupied by an old abandoned construction (which left almost no green footprint of any soil absorption), and in a neighborhood that required it due to its proximity to the Vega stream; the decision was made to release half of the lot, even though it was a very small lot for the development of a high-rise building, to begin with. Leaving the largest amount of absorbent land surface possible, with a mixture of green plants and adding new vegetation with leafy trees that help clean the surrounding air.

Escaping the monotony and the norm, providing a greater confluence of inhabitants of different social classes and natures, cooperating to converge and form a community in which everyone can interact. Each of the apartments has cross ventilation, direct or indirect entry of the sunlight, and permanent contact with the vegetation of the building both on the inside as well as on the outside, with its large terrace balconies. Each and every one of the units is supported by a large outdoor space with its own public vegetation provided by the building, where the large trees of the interior courtyard and those in the background embrace and contain the entire neighborhood. The building places great emphasis on sensory perceptions, as a primitive and inquisitive matter of mere being. Making a huge commitment to its large open spaces, allowing a full experience with nature. A sensation that is so basic and fundamental, but at the same time so scarce and forgotten in the great metropolises and in this case in our Buenos Aires. We always say that "The project strengthened the building"; providing it with new species, which absorb CO_2 and produce the oxygen necessary to improve the life of the building and of the city itself. The RH+ collective housing building gives those who inhabit it a ray of sunlight. Perhaps it bounces off a leaf of a tree in a delighting play of light and shadow, avoking tranquility that passes through a peaceful, calm, and serene time in coexistence with neighbors.

Location Saavedra, Buenos Aires, Argentina **Use** Apartment **Gross floor area** 438m² **Completion** 2022 **Design team** Marcelo Rebecca, Luciano Rebecca, Matias Rebecca, Victoria Polleti **Photographer** Luis Barandiaran

알에이치플러스 프로젝트는 도시의 단점을 극복하고 거주자의 삶의 질을 향상시키기 위해 시작되었다. 토양 내 녹색 발자국이 거의 남지 않은 오래된 폐건물이 95%를 차지하는 부지는 베가 하천과의 근접성 으로 인해 주민의 삶을 위해서 개발이 절실했으므로 작은 부지에 초고층 건물을 올리는 것이었음에도 처음부터 부지 절반을 오픈하기로 하였다. 흡수 가능한 지표면을 가능한 많이 남기고 주변 공기 정화를 위해 녹색 식물에 잎이 무성한 식목을 더했다.

다양한 계층 및 성격을 가진 거주민들이 경직성과 규범에서 벗어나 서로 자유롭게 소통하는 공동체를 형성할 수 있는 공간을 제공하고자 하였다. 각 아파트 공간은 교차 환기, 직간접 채광, 식물을 통해 유기적으로 연결되는 내부와 외부 및 대형 테라스 발코니가 특징이다. 모든 개별 가구에는 건물에서 조경을 제공하는 넓은 야외 공간이 조성되며, 큰 나무들이 심겨진 내부 안뜰과 마당이 거주민들에게 공유된다.

이 건축물은 단순한 존재의 본능과 인지를 통한 감각적 인식에 중점을 두었다. 넓은 열린 공간을 조성함으로써 자연과의 일체감을 경험하도록 했다. 이는 매우 본능적인 감각이지만 대도시나 이곳 부에노스아이레스에서는 찾아보기 어려운 잊혀진 감각이다. '빌딩을 강화 시키는 프로젝트'를 지향하며 건물은 물론 도시의 수명을 향상시키는 데 필요한 산소를 만들어 내고 이산화탄소를 흡수하는 조경에 중점을 두었다. 알에이치플러스 빌딩은 거주민들에게 자연 채광을 선물한다. 아마도 그것은 빛과 그림자의 유희 가운데 나무 잎사귀에 튕겨 나가며 이웃과 공존하는 고요하고 편안하고 평화로운 공간을 만들어 낼 것이다.

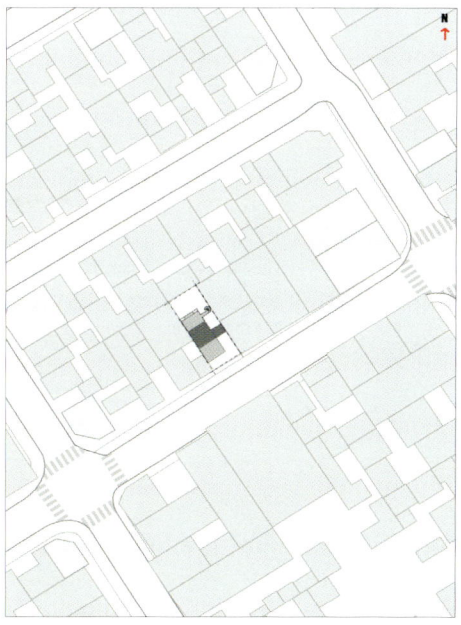

SITE PLAN

← Front facade

Corner view
Night view

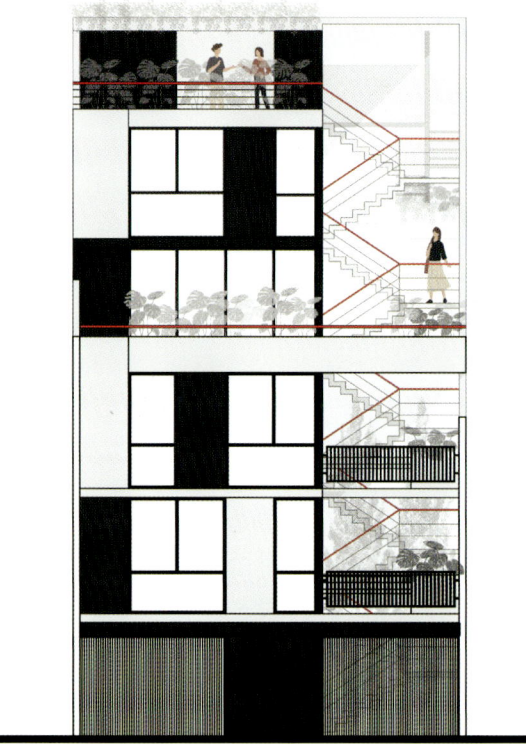

FRONT ELEVATION

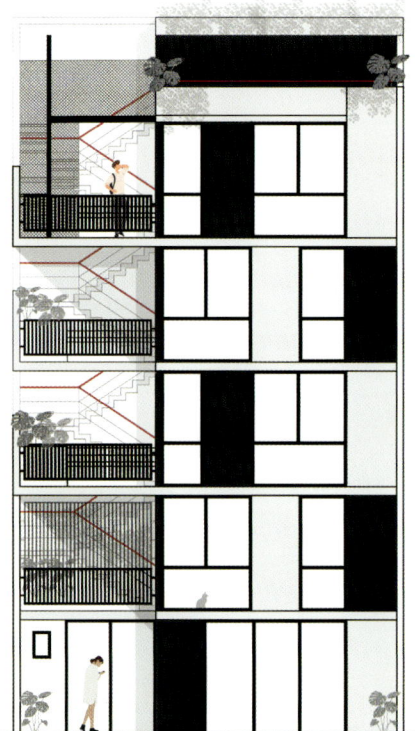

REAR ELEVATION

Vegetation

Sunshine

Neighborhood interaction

Crossed ventilation

DIAGRAM

← Entrance → Courtyard

← Exterior stair

SECTION

↑ Bicycle parking lot   ↙ Entrance of unit   ↘ Interior courtyard

← Interior view   → Interior view

1 MAIN ENTRANCE	5 LIVING ROOM	9 OUTDOOR TERRACE	13 ROOFTOP TERRACE
2 STORAGE	6 ENTRANCE HALL	10 VERANDA	14 RESTROOM
3 EXTERIOR STAIRCASE	7 BEDROOM	11 COURTYARD	
4 DINING & KITCHEN	8 BATHROOM	12 ROOFTOP KITCHEN	

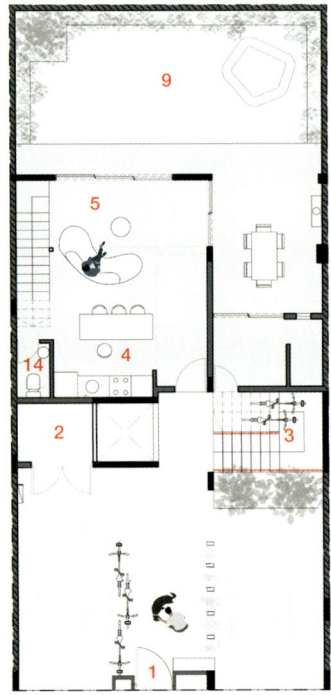
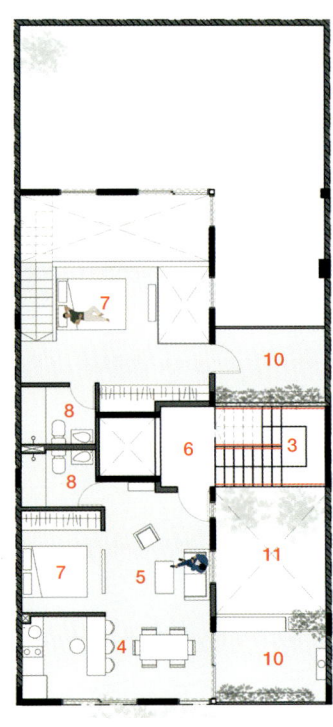
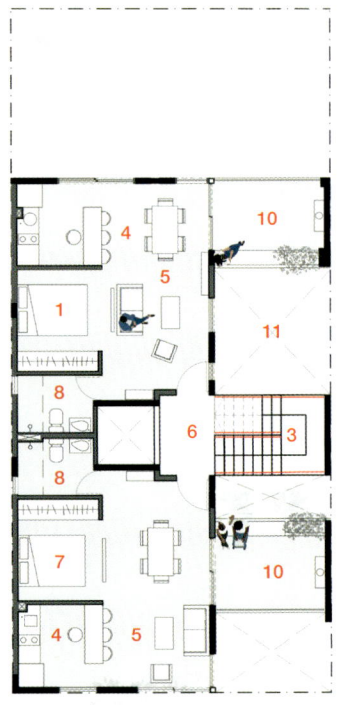

GROUND FLOOR PLAN   **1ST FLOOR PLAN**   **2ND FLOOR PLAN**

← Courtyard   → Terrace

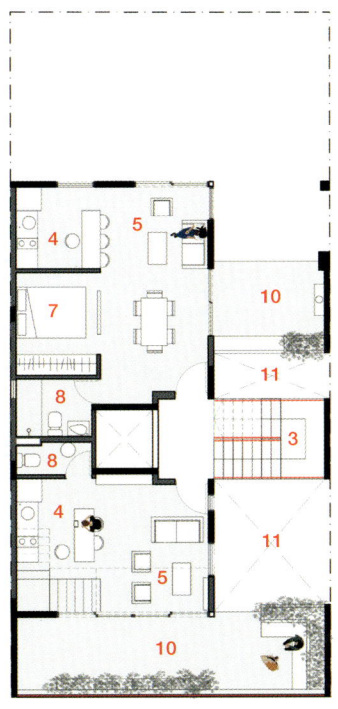

3RD FLOOR PLAN

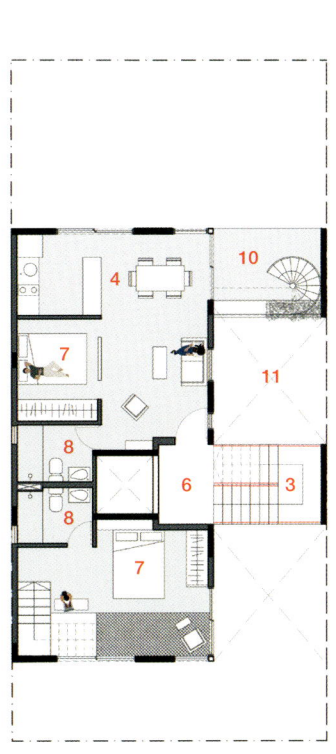

4TH FLOOR PLAN

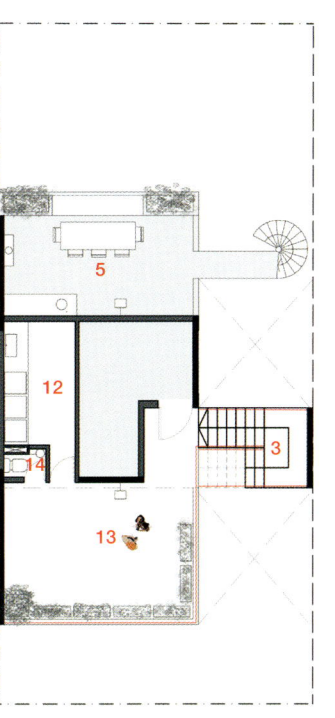

ROOF FLOOR PLAN

THE HOUSE IS LOCATED ON A PLOT WITH A STEEP SLOPE in the Barranco de Soterrana, a natural watercourse near Valencia. An industrialized, fragmented and unplugged house idea developed, integrating passive systems and self-sufficient energy generation. In this mid-hill location, the house, designed to be manufactured with industrialized components in three months and then installed in a few days, is accommodated on a light metallic structure that allows habitability in this complex topography, without manipulating or taming it.

Therefore, a sequence of elevated domestic, interior and exterior spaces is developed, thus maintaining the ground runoff regime, and allowing future extensions, incorporating other modules, or simply colonizing parts of this hybrid structural framework.

The house explores this fragmentation, consistent with the proposed construction systems, generating a profound daily relationship with the landscape, without resorting to large glazed surfaces or other devices incompatible with the climate and the budget. Consequently, all parts of the house in this sequence of spaces have direct access to the outside and play with the games of the proposed structure.

The image is only part of much greater complexity.

Location Barranco de Soterraña, Valencia, Spain **Use** Housing **Plot surface** 1.113 m² **Constructed area of the house** 142 m² **Completion** 2021 **Collaborating architects** Guillermo Gomez, Alvaro Arroyo, Zuzanna Cieslewicz **Structures**: Daniel Carratala **Facilities** Ingenet, S.L. **Quantity Surveyor** Miguel Monteagudo Cuevas **Photographer** Diego Opazo

산업화 하우스 rNrH
INDUSTRIALIZED HOUSE rNrH

ARCHITECT : JUAN MARCO ARQUITECTOS / JUAN MARCO MARCO

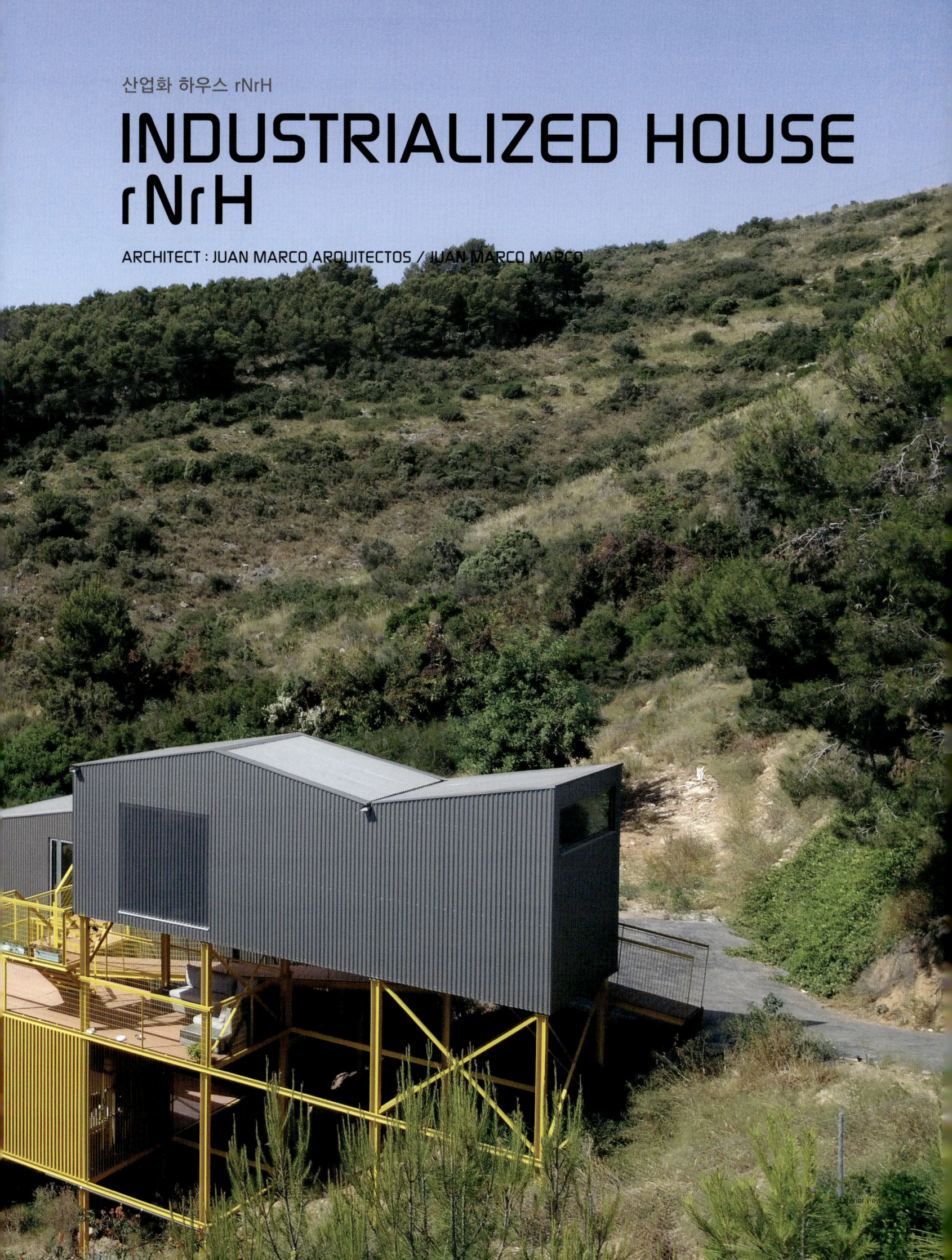

Exterior view

↑ Light metallic structure house

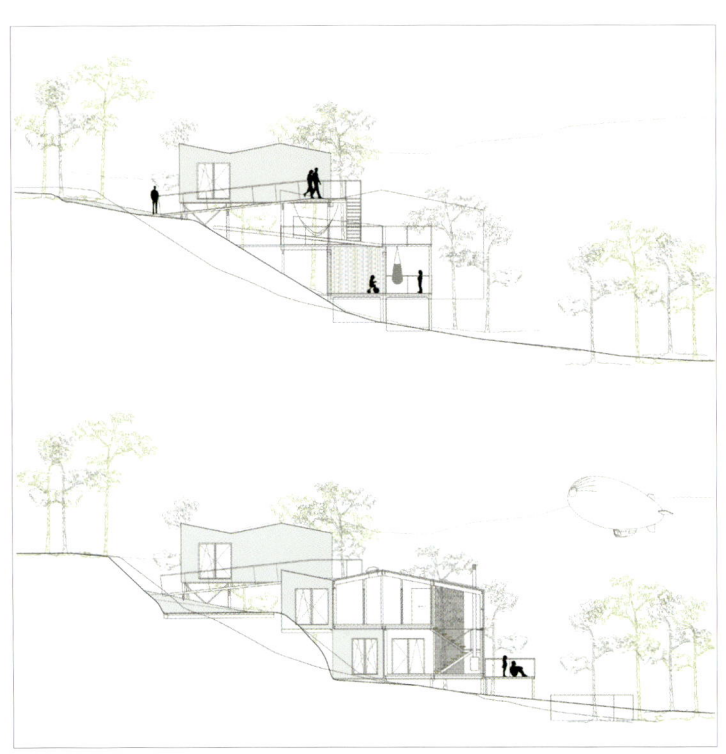
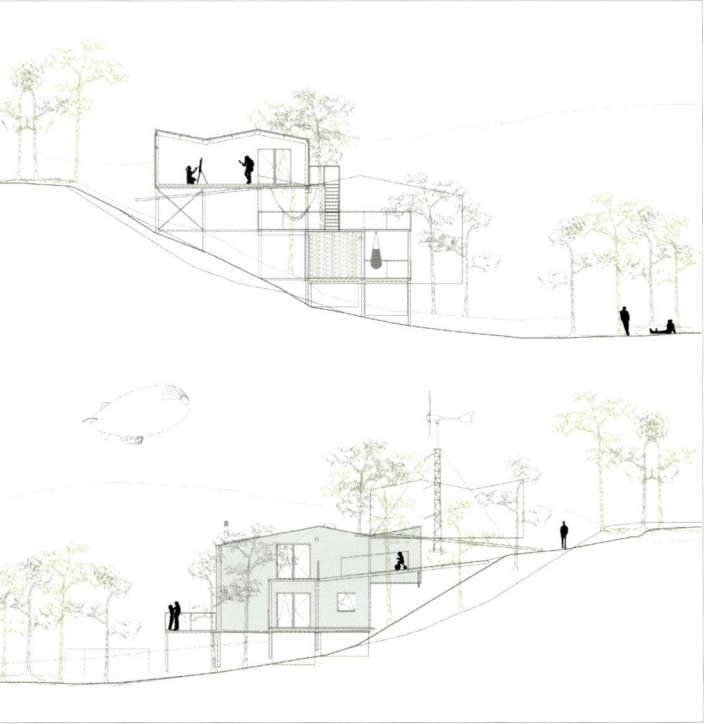

ELEVATION & SECTION

↱ Light metallic structure house
↳ Hybrid structural framework

발렌시아 근처의 자연 수로인 바랑코 드 소테라나의 가파른 경사에 위치한 주택이다. 전기를 사용하지 않는 단절된 구조의 산업화 주택에 대한 아이디어는 수동 및 에너지 자급 통합 시스템으로 이어져 있다.

중간 언덕에 위치한 이 주택은 3개월간 산업 부품으로 완성해 며칠 안에 설치하도록 설계되어 별도의 조작이나 개조 없이도 복잡한 지형에서 활용될 수 있는 가벼운 금속 구조로 이루어져 있다.

그로 인해 지상층의 흐름이 그대로 유지되면서 다른 모듈과 통합해 확장되거나 단순히 하이브리드 구조의 부품으로 이루어진 일련의 내외부 공간이 구성되어 있다.

이런 건축 방식과 조응하며 단절된 구조를 이루고 있는 주택은 기후와 예산에 맞지 않는 대형 유리 표면이나 기타 장치 없이도 자연과 어우러진 공간을 이루고 있다. 결과적으로 이 일련의 주택 공간은 외부에서 직접 접근할 수 있는 독특한 구조로 형성되었다. 이 이미지는 훨씬 복잡한 구조의 일부일 뿐이다.

↑ Installed systems in a few days   ↙ Future extensions systems

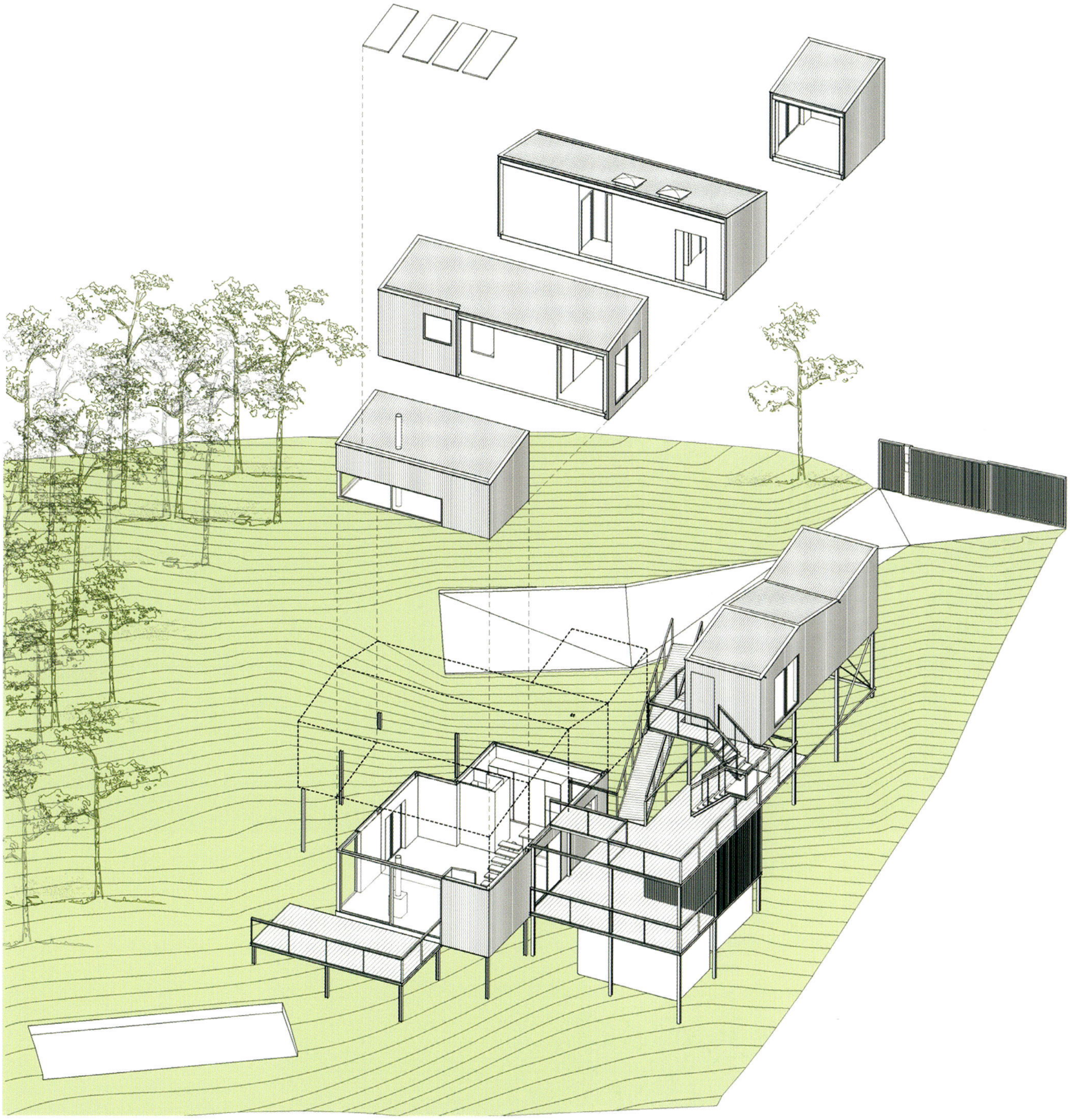

EXPLODED VIEW

← Modeling   → The house located on a plot with a steep slope

↑ House have direct access to the outside   ↵ Exterior night view

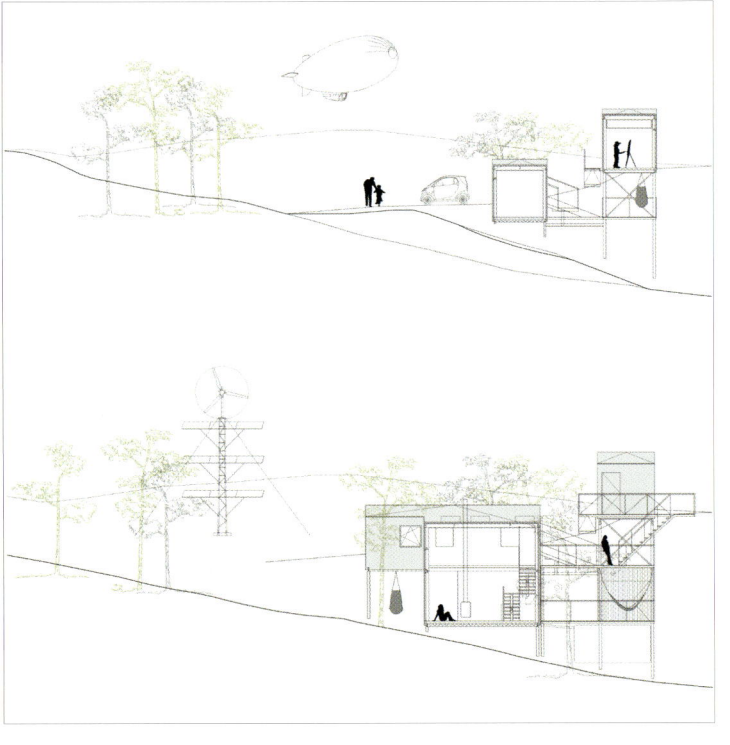

SECTION

↑ Overall view of the building   ↵ House have direct access to the outside & play   ↳ Hybrid structural framework

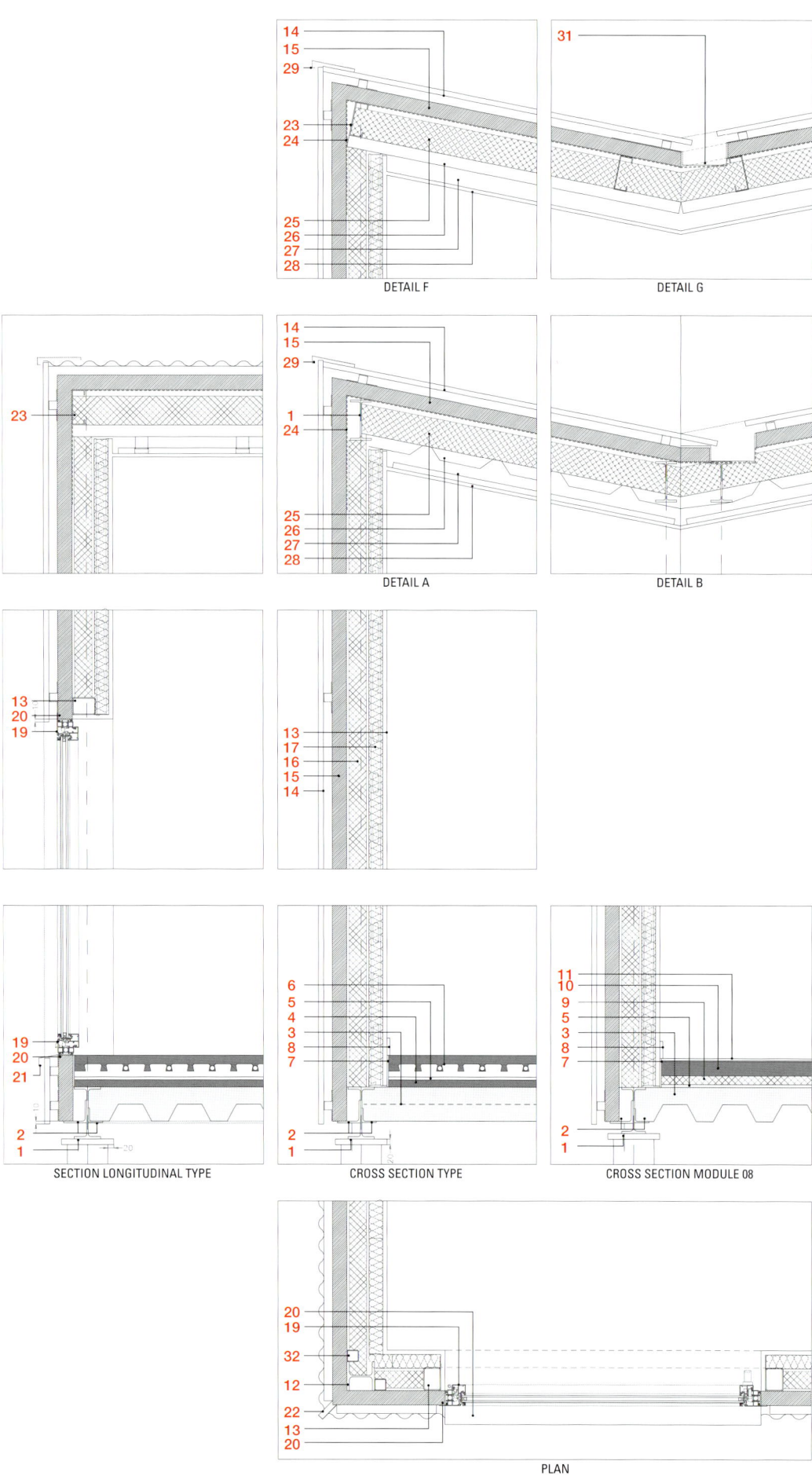

CONSTRUCTIVE SECTION DETAIL

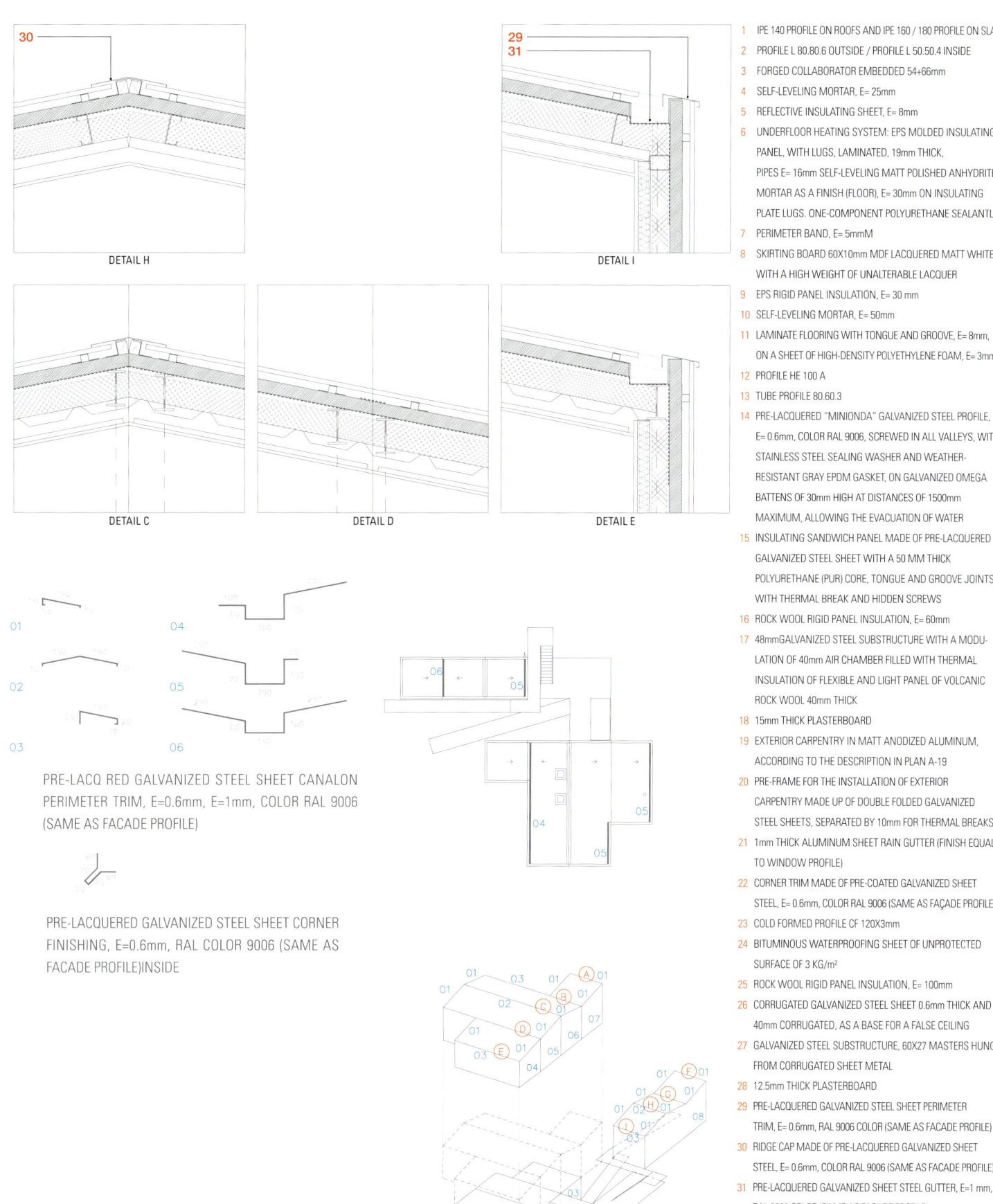

1. IPE 140 PROFILE ON ROOFS AND IPE 160 / 180 PROFILE ON SLABS
2. PROFILE L 80.80.6 OUTSIDE / PROFILE L 50.50.4 INSIDE
3. FORGED COLLABORATOR EMBEDDED 54+66mm
4. SELF-LEVELING MORTAR, E= 25mm
5. REFLECTIVE INSULATING SHEET, E= 8mm
6. UNDERFLOOR HEATING SYSTEM: EPS MOLDED INSULATING PANEL, WITH LUGS, LAMINATED, 19mm THICK, PIPES E= 16mm SELF-LEVELING MATT POLISHED ANHYDRITE MORTAR AS A FINISH (FLOOR), E= 30mm ON INSULATING PLATE LUGS. ONE-COMPONENT POLYURETHANE SEALANTL
7. PERIMETER BAND, E= 5mmM
8. SKIRTING BOARD 60X10mm MDF LACQUERED MATT WHITE WITH A HIGH WEIGHT OF UNALTERABLE LACQUER
9. EPS RIGID PANEL INSULATION, E= 30 mm
10. SELF-LEVELING MORTAR, E= 50mm
11. LAMINATE FLOORING WITH TONGUE AND GROOVE, E= 8mm, ON A SHEET OF HIGH-DENSITY POLYETHYLENE FOAM, E= 3mm
12. PROFILE HE 100 A
13. TUBE PROFILE 80.60.3
14. PRE-LACQUERED "MINIONDA" GALVANIZED STEEL PROFILE, E= 0.6mm, COLOR RAL 9006, SCREWED IN ALL VALLEYS, WITH STAINLESS STEEL SEALING WASHER AND WEATHER-RESISTANT GRAY EPDM GASKET, ON GALVANIZED OMEGA BATTENS OF 30mm HIGH AT DISTANCES OF 1500mm MAXIMUM, ALLOWING THE EVACUATION OF WATER
15. INSULATING SANDWICH PANEL MADE OF PRE-LACQUERED GALVANIZED STEEL SHEET WITH A 50 MM THICK POLYURETHANE (PUR) CORE, TONGUE AND GROOVE JOINTS WITH THERMAL BREAK AND HIDDEN SCREWS
16. ROCK WOOL RIGID PANEL INSULATION, E= 60mm
17. 48mmGALVANIZED STEEL SUBSTRUCTURE WITH A MODULATION OF 40mm AIR CHAMBER FILLED WITH THERMAL INSULATION OF FLEXIBLE AND LIGHT PANEL OF VOLCANIC ROCK WOOL 40mm THICK
18. 15mm THICK PLASTERBOARD
19. EXTERIOR CARPENTRY IN MATT ANODIZED ALUMINUM, ACCORDING TO THE DESCRIPTION IN PLAN A-19
20. PRE-FRAME FOR THE INSTALLATION OF EXTERIOR CARPENTRY MADE UP OF DOUBLE FOLDED GALVANIZED STEEL SHEETS, SEPARATED BY 10mm FOR THERMAL BREAKS.
21. 1mm THICK ALUMINUM SHEET RAIN GUTTER (FINISH EQUAL TO WINDOW PROFILE)
22. CORNER TRIM MADE OF PRE-COATED GALVANIZED SHEET STEEL, E= 0.6mm, COLOR RAL 9006 (SAME AS FAÇADE PROFILE)
23. COLD FORMED PROFILE CF 120X3mm
24. BITUMINOUS WATERPROOFING SHEET OF UNPROTECTED SURFACE OF 3 KG/m²
25. ROCK WOOL RIGID PANEL INSULATION, E= 100mm
26. CORRUGATED GALVANIZED STEEL SHEET 0.6mm THICK AND 40mm CORRUGATED, AS A BASE FOR A FALSE CEILING
27. GALVANIZED STEEL SUBSTRUCTURE, 60X27 MASTERS HUNG FROM CORRUGATED SHEET METAL
28. 12.5mm THICK PLASTERBOARD
29. PRE-LACQUERED GALVANIZED STEEL SHEET PERIMETER TRIM, E= 0.6mm, RAL 9006 COLOR (SAME AS FACADE PROFILE)
30. RIDGE CAP MADE OF PRE-LACQUERED GALVANIZED SHEET STEEL, E= 0.6mm, COLOR RAL 9006 (SAME AS FACADE PROFILE)
31. PRE-LACQUERED GALVANIZED SHEET STEEL GUTTER, E=1 mm, RAL 9006 COLOR (SAME AS FACADE PROFILE)
32. TUBULAR PROFILE 40.40.2 FOR FIXING INSULATING SANDWICH PANEL

CONSTRUCTIVE SECTION DETAIL

↑ Interior view in the house

↑ Assembly completion process view

PROFILE

↘ page 004

Nguyen Dang Tuong | NDT Architecture
He was born in 1991 in Bac Ninh City, Viet Nam. He is architect and director of NDT Architecture JSC in Viet Nam. He is now living and working in Bac Ninh city of Viet Nam. He established NDT Architects since 2017 with mission bring green architecture for everyone in Vietnam. They often focus on housing design with the desire to improve living space for everyone. The message they want to convey through this project is "Respect and protect the nature." Hopefully through this project they can somehow inspire other architects in their community to follow our path - green architecture.

↘ page 014, 234

Shikwan Yang, Tomonori Miura, Tatsuhito Ono | T2P Architects office & P.O.T lab
Planning, designing, and supervision of construction, Interior design and supervision of housing,shops, offices, exhibition, events, and so on Planning, designing, and manufacturing of furniture and product arts Creation of master plans and consulting services for communities, environment, and facilities Education, lectures, and publishing with regard to urban planning and architecture Overseas expansion of the above business.

↘ page 026

Kensuke Aisaka | Aisaka Architects' Atelier
He is a first-class architect of Aisaka Architects' Atelier, born in Tokyo, Japan. After graduating from the Department of Architecture at the University of Tokyo, he worked at Tadao Ando Architects & Associates from 1996 to 2002. He founded Aisaka Architects' Atelier in 2003. He is also a lecturer at Toyo University and Hosei University. He is a member of the Japan Institute of Architects, Tokyo Society of Architects & Building Engineers and Architectural Institute of Japan.

↘ page 034

Eloisa Ramos, Moreno Castellano | Ramos Castellano Arquitectos
The studio, is one of the first african based architecture studio, with projects published worldwide. Since the first projects the main purpose of RamosCastellano arquitectos was to use architecture and materia, to change consciences. Each work is a mixture between Art, Nature, Social contest, local knowledge, Biology and Technology. The studio with an international team, based in Mindelo, Cabo Verde, West Africa, is born from the collaboration between:
Eloisa Ramos, Caboverdian architect with an approach to architecture visionary, direct, clear, and simple. Moreno Castellano, Mediterranean multimedia artist triyng to use architecture and art, as a social revolutionary tool.

↘ page 046, 194

Inkeun Ryoo, Doran Kim, Sangkyong Jeong | YOAP architects Ltd.
YOAP architects is a group planning, designing, marketing; architecture/ interior/ items. We aim not to limit architectural ideas to buildings but to extend it to diverse design areas. Seeking for sustainable enjoyments in design works, we also hope to share it. As its name, YOAP ("nearby" in Korean) will hold an image of friendly neighbor.

↘ page 058

Ricardo Dias, Bruno Tinoco | Falanstério atelier de arquitectura
Embracing new challenges with creative solutions, professional knowledge and passion. Falanstério – architecture studio, was established in 1996 by Bruno Tinoco and Ricardo Dias. Working in architectural and interior design projects, Falanstério joins today a group of qualified professionals and establishes business partnerships in order to provide a better response to each project. Due to the professional experiences of its partners this studio is qualified to give an overall support to commercial projects. Falanstério presents sustainable solutions in every project, refurbish or new ones.

↘ page 068

Ippolito Pestellini Laparelli | 2050+
2050+ is an interdisciplinary agency based in Milan, working across technology, the environment, politics and design and visual culture. 2050+ operates in different fields and through different media, engaging meaningfully with projects spanning across architecture, installation design, curatorial and research practices. 2050+ uses intersectionality as a framework to look at contemporary world dynamics and issues and works transversely across commercial and non-commercial domains. 2050+'s work has been featured on several editorial platforms, including FlashArt, e-flux, Volume, among others.

↘ page 078, 136

Hideo Kumaki | Organic Design Inc.
Organic Design Inc. is a Tokyo-based architecture firm that is a creative group of architects and interior designers who struggle to create dynamism in space with an emphasis on curves. Our office has won the Grand Prix in Japan(SDA Award), as well as national and international awards in USA, Germany, Hong Kong, and other countries.
We have confidence on providing the best solutions and design value for any range of work, including architecture, interiors, landscape, and sign, whatever you need.

↘ page 088

George Loizou | Loizou Architects + Associate
As architects we believe in design process and continuous development. Each project is treated as a design experimentation opportunity. The design methodology combines thorough analysis of the brief, interpretation of existing site conditions and minimal design intervention. Sustainability and pragmatism of spaces characterise the design hierarchy. Over the years, the practice has undertaken a variety of projects. The team has engaged with different scale projects ranging from residential to commercial and industrial. Our clientele includes individuals as well as companies. As a team we are able to deal with high and low budget projects. We aim to present our work through friendly and attractive display means. However, the practice's objectives and subsequent approach towards architecture remain the focal principles.

↘ page 098

Abimantra Pradhana, Osrithalita Gabriela | AGo Architects
AGo Architects is an award-winning Indonesian architecture consulting firm based in Jakarta. We believe in architecture practices that consider its impacts on the environment and the society it lives in. This design thinking ensures that architecture is not merely a physical experience, but rather an aesthetic interpretation of the harmonic relationship between humans and their physical environment. We strive to deliver functional, sustainable, and socially responsible architecture by carefully curating its raw materials, efficient design, and inclusivity in mind.

↘ page 110

Juan Antonio Santa-Cruz Garcia | Santa-Cruz Arquitectura
Santa-Cruz Arquitectura is an architecture studio based in Murcia, Spain. They understand architecture as a discipline that contributes to the sustainable development of society and respect for nature and our planet. To do this, they address the design process from concepts such as biophilia, bioclimatism, equity, archaeology, phenomenology or crafts.
They have been internationally awarded by the Ibero-American Design Biennial or the German Design Awards and nationally by the Regional Architecture Awards of the Region of Murcia.

↘ page 122

Clara Murado, Juan Elvira | Murado & Elvira Arquitectos
Murado & Elvira is a Madrid-based multidisciplinary office dedicated to innovative architecture and interior design. They have been finalists at the Norwegian National Architecture Prize in 2012 and the Mies van der Rohe Award in 2019. They design relaxed and airy environments that allow for an immediate personal appropriation on the part of the user. They propose a sustainable approach to design, where user participation, shared creativity and building techniques seek to benefit both society and nature.

↘ page 144

Dennis Mueller, Matthias Siegert | VON M
VON M is led by the two partners Matthias Siegert and Dennis Mueller. Starting point is always an intensive examination of the qualities and characteristic features of the respective environment. Solutions are created closely linked to the context, both conceptually and in terms of their materialisation, which are coherent and comprehensible in themselves. In addition to several residential buildings, the main tasks in recent years have included projects for cultural institutions and public clients, as well as numerous existing buildings.

↘ page 152

Raul Sanchez | Raul Sanchez architects
Raul Sanchez architects is an award winning architecture and design practice based in barcelona, founded by Raúl Sánchez.
Raúl Sánchez is graduated architect from the architecture school in Granada, Spain. Since 2005 resides and works in Barcelona developing a professional activity which escapes specialization in order to cover all types of work and projects related to architecture, interiorism, urbanism and design. They firmly believe in architecture to transform not just a space, but the life lived in such space. He is currently professor in 'private perimeters', a postgraduate diploma in the school Elisava, Barcelona. Regardless of the type, scale, location or budget, every new work is a new challenge.

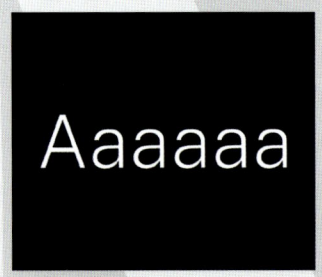
↘ page 164

Triet Le, Ho Ngoc Nhung | Aaaaaa
6A is an independent architecture atelier that pushes the boundary of contemporary local architectural industry. Our process involves meticulous exploration of forms, function, and design language that respond to constraints of brief, budget, site condition and constructability.

↘ page 172

Gonzalo Pardo | gon architects
gon is a Madrid-based architecture and design office headed by Gonzalo Pardo since 2014. Its practice focuses on research and development of singular architectural projects of different scales ranging from urban planning to buildings to interior construction. The common denominator of their works is a playful, experimental, critical, and optimistic view of the contemporary. In a constant dialogue based on observation and details, their interest focuses on the creative processes of architectural design and construction, as well as the role of mediation and communication of architecture.

↳ page 180

Rem Koolhaas, David Gianotten | OMA
Rem Koolhaas (Rotterdam, 1944) founded OMA in 1975 together with Elia and Zoe Zenghelis and Madelon Vriesendorp. He graduated from the Architectural Association in London and in 1978 published Delirious New York: A Retroactive Manifesto for Manhattan. In 1995, his book S,M,L,XL summarized the work of OMA in "a novel about architecture".
David Gianotten is the Managing Partner and Architect of OMA. David oversees the overall organizational and financial management, business strategy, and growth of OMA in all markets, in addition to his own architectural portfolio. David currently leads the design and construction of projects in different regions, including Amsterdam's Bajes Kwartier. David joined OMA in 2008, launched OMA's Hong Kong office in 2009, and became partner in 2010.

↳ page 204

Noppachai Akayapisud, Sathika Jienjaroonsri | Spacecraft Co., Ltd.
Spacecraft Co., Ltd. is officially found in 2020. However Noppachai and Sathika have been working together on projects since 2016. With 10 years of combined experiences in design fields, we have passion in designing and challenging projects with board ranges of clients and types of projects. Throwing new ideas, concept and exploring new materials are what we love to do. Spacecraft always offer new perspectives and visuals to clients with forward-thinking ideas. We see clients as our partners and always work closely with them to achieve design and projects' goals. Now we are a small design studio based in Bangkok but full of young, strong and passionate interior designers and architects that we are very proud of.

↳ page 210

Alix Héaume, Adrien Robain | rh+ architecture
The agency, based in Paris, was founded in 2000 by Alix Héaume and Adrien Robain. Since the beginning the agency has been working on many projects of different type and size, in places and with multiple challenges like, for example, the requalification of an ancient barracks in the center of Paris, Guyana's new housing in their first eco-district or the construction of an intergenerational residence in Paris. This diversity is fundamental for the agency, which has a desire for constant renewal, discovery, learning and design of tailor-made projects that fit their territory.

↘ page 218

Tan Yamanouchi | Tan Yamanouchi & AWGL

Born in Sapporo, Hokkaido. From an early age, he learned practical architectural design at his father's office and site, an architect. Completed the master's course at Keio University Graduate School. Joined Hakuhodo Inc. as a Market Design (MD). With a vision of "More than an Architect," he proposes a new style of work as an architect, from architectural design to branding and advertising PR, through concrete examples. The architectural design is based on Tan Yamanouchi's unique architectural philosophy, "Architecture of Ghost," which is derived from the ancient East Asian thought that "Gods (souls) reside in all people, things, animals, and plants." 2017 , Awarded "Architects of the Year 2017" from the Architectural Institute of Japan. Won the title.The rarity and ability to carry out both architectural design and branding in one go are highly evaluated internationally.

↘ page 226

Han Kuo | Supra-Simplicities

Supra-Simplicities is an architectural studio, led by Han Kuo, with teams based in Taiwan and New York. Han Kuo, a licensed architect in Taiwan, brings over 7 years of experience working and collaboration with renowned offices including OMA, 3XN, and Kengo Kuma. At OMA, he played a key role in supervising the construction of the Taipei Performing Arts Center. He earned his degree from Columbia GSAPP and holds a position as visiting assistant professor at Pratt Institute.

↘ page 242

Sivichai Udomvoranun | AOMO

AOMO (Architecture of My Own) Architectural firm based in Bangkok, Thailand. They design every building types. AMOMO team focuses on the process to ensure that the final results are nothing less than perfect spaces for our clients. They carefully design from simple to sophisticated details that will last through time and make sure that clients' budget is carefully spent. They try to create something new for our clients since they understand that no one wants to be like others.

↘ page 252

Javier Goldenberg, Matias Goldenberg | HERMANOS GOLDENBERG

Javier Goldenberg (1964) : Architect at the University of Buenos Aires in 1988. Master MDI IV Catholic University of Buenos Aires 2001. Postgraduate Specialization in Restoration of Historic Buildings at the Catholic University of Buenos Aires 2005.

His projects are aimed at adding value, refunctioning and remodeling buildings, old, patrimonial and historical buildings, houses and apartments and new medium-scale works with cutting-edge technology.

Mathias Goldenberg (1972) : An invitation from his father to travel to Entre Rios to visit a labor engineer awakens the passion for construction inherited from his family. In 1998, after spending a year with his father, he began studying an MBA in building development. The following year, with his brother Javier, they developed a tower crane management and assembly venture, thus paving the way for the use of this technology in the country, and thanks to the experience acquired in the United States, they created the first fully automated events company. generating a change in the concept of logistics in this sector.

↘ page 260

Marcelo Rebecca, Luciano Rebecca, Matias Rebecca, Victoria Polleti | RBK arquitectura

We are an architecture studio made up of professionals from all areas. We conceive architecture as the result of an active, versatile and enriching process for the entire work team.

This approach allows us to obtain as a result, works that work and expand the expectations of our clients.

↘ page 268

Juan Marco Marco | Juan Marco Arquitectos

Juan Marco Arquitectos is an architecture office located in Valencia, with a vocation of laboratory and place of study. They ara very involved in facing the theory and practice of the construction of architecture and more concerned with the economic, technological, social, and environmental considerations of their work than with signing calligraphies.

The founder, Juan Marco Marco, studied architecture at the Polytechnic University of Valencia and the University of Buenos Aires. From 2001 to 2003 he collaborated with the architect Josep Llinás, in Barcelona. In 2004 he founded Juan Marco Marco architecture studio, since 2015: Juan Marco Arquitectos.

Publisher | Heungchae Jung
Editorial Dept. | Joonyong Jung, Eunjae Ma
Design Dept. | A&C design team

Print in Korea
ISBN | 978-89-7212-000-1
Price | USD 68 (68,000won)
Registration No. 2004-000166

© **A&C Publishing**
9F, 15, Teheran-ro 22-gil, Gangnam-gu, Seoul, Korea
T: +82-2-538-7333
www.ancbook.com

Copyright A&C Publishing and may not be
reproduced in any manner or from without permission.